GUIDELINES TO HYDRAULIC TRANSIENT ANALYSIS

Guidelines to Hydraulic Transient Analysis

S Pejović, A P Boldy
and D Obradović

Technical Press

Published by
Gower Technical Press Ltd.,
Gower House,
Croft Road,
Aldershot,
Hants. GU11 3HR,
England.

Gower Publishing Company,
Old Post Road,
Brookfield,
Vermont 05036,
U.S.A.

British Library Cataloguing in Publication Data

Pejović, S.
Guidelines to hydraulic transient analysis.
1. Hydraulic machinery 2. Water hammer
3. Water–power electric plants
I. Title II. Boldy, A.P. III. Obradović, D.
621.31'2134 TJ844

Library of Congress Cataloguing in Publication Data

Pejović, S.
Guidelines to hydraulic transient analysis.

Bibliography: p.
Includes index.
1. Hydraulic transient. 2. Hydroelectric power plants – Safety measures.
I. Boldy, A.P., II. Obradović, D. III. Title.
TC174.P36 1987 621.31'2134 86–31925

ISBN 0 291 39723 9

Printed and bound in Great Britain by Dotesios(Printers)Limited,
Bradford-on-Avon,Wiltshire

Preface

The aim of this book is to assist engineers who are responsible for the design, construction, commissioning and operation of hydroelectric power plants; it does not explain the philosophy of transient flow or present methods for the analysis of such phenomena. The book is intended as a handbook, assisting engineers to determine the main parameters for a proposed installation, to define the necessary data for, and scope of, the transient analysis, to select the most dangerous cases and to draw the correct conclusions.

The information presented in this book is important for large and small hydro power plants and pumping systems. Analysis of hydraulic transients and hydraulic vibrations is a very time consuming and expensive activity. It is therefore important in the case of small power plants that only the most critical transients are selected and analysed. Unfortunately, it is quite impossible to prescribe rules which will apply in every conceivable case, at least for the time being. Therefore the analysis of transients and vibrations for small power plants should be given to the experienced engineers.

The potential dangers for hydro plants due to transient phenomena are discussed in Chapter 1 together with the principles, and devices for pressure surge and hydraulic vibration protection. (Some accidents resulting from transient phenomena is included to show the significance and potential dangers for hydro plants).

The main elements of hydro power plants, for the transient flow analysis, are described in Chapter 2. Characterisitic steady state regimes are defined. Characteristic parameters of transient flow analysis are explained in Chapter 3.

Chapter 4 contains a brief description of turbine governors with important parameters listed. Various regulating devices are mentioned, and the opening—up closing—down modes are explained in detail, stressing the important parameters (maximum seed of servomotor piston, for instance).

Chapter 5 deals with input data necessary for the analysis of transient regimes in hydro power plants. The extent of investigations on a small—scale model is specified in some cases. Other data, needed for the analysis on the level of feasibility study and/or general design, are also listed there.

Various transient regimes are classified in Chapter 6. There are normal cases which the plant has to withstand without difficulties, then emergency cases, when no substantial damage should occur in the plant, and lastly catastrophic cases when some damage cannot be prevented.

Computational runs which should be performed in the analysis are listed in Chapter 7. Parameters of the analysis are discussed and the minimum output specified.

Measurements and investigations performed during the various phases are discussed in Chapter 8.

The recommended range of main parameters are discussed in Chapter 9. Some guidelines for protection against water—hammer are also given in Chapter 10.

A well—performed analysis of transient regimes might be very fruitful: it can ensure the safety of the plant, eliminate or attenuate operational problems, and prevent serious damage. The authors do hope that this book will help other engineers to achieve their ends.

The authors welcome comments, opinions, and suggestions from colleagues regarding the material presented in this book. All contributions will be carefully evaluated and where appropriate incorporated in the next edition. Please address any contributions to:

1. Professor Stanislav Pejovic,
Faculty of Mechanical Engineering,
University of Belgrade,
11000 Beograd, 27 Marta 20,
Yugoslavia.

2. Dr. Adrian P. Boldy,
Department of Engineering,
University of Warwick,
Coventry CV4 7AL,
England.

Contents

Illustrations

Notation

A	m^2	flow area characteristic area of hydraulic machine ($\pi D^2/4$)
A_t	m^2	cross–sectional area of tunnel
A_{st}	m^2	cross–sectional area of surge tank
A_r	m^2	effective tunnel area
a	ms^{-1}	velocity of pressure wave (wavespeed)
a_o	m	wicket gates opening
b_p	–	permanent speed droop
c	ms^{-1}	flow velocity
c_o	ms^{-1}	initial flow velocity
D	m	characteristic diameter of hydraulic machine runner
F_a	N	axial hydraulic thrust
F_{11}	–	unit axial hydraulic thrust (Fa/D^2H)
f	Hz	frequency
f_{th}	Hz	theoretical natural frequency
g	ms^{-2}	gravitational acceleration
H	m	head
H_o	m	rated head
H_L	m	head loss
h	–	dimensionless head (H/Ho)
I	kgm^2	inertia of rotating parts ($MD^2/4$)

K	–	valve flow coefficient
K_{st}	$m^{-5}s^{2}$	surge tank throttle flow coefficient
L	m	length of a tunnel or penstock
L_e	m	equivalent length of a tunnel or penstock
M	Nm	torque
M_o	Nm	rated torque
M_{11}	–	unit torque (M/D^3H)
m	–	dimensionless torque (M/M_o)
MD^2	kgm^2	polar moment of inertia of rotating parts (4I)
n	min^{-1}	speed of rotation
n_o	min^{-1}	rated speed of rotation
n_{11}	–	unit speed ($nD/\surd H$)
P	W	power
P_g	W	generator power
P_h	W	hydraulic power
P_ℓ	W	power loss
P_m	W	motor input power
p	Pa	pressure
p_s	Pa	design pressure for piston
Q	m^3s^{-1}	volumatic flow rate (discharge)
Q_o	m^3s^{-1}	steady state volumetric flow rate
Q_{11}	–	unit discharge ($Q/D^2\surd H$)
T_a	s	unit starting time
T_f	s	servomotor closing time corresponding to maximum servomotor velocity
T_f^H	s	reduced servomotor closure time
T_f^ω	s	extended servomotor closure time
T_{fo}	s	minimum opening time of servomotor
T_h	s	time for decelerated closure of wicket gates
T_q	s	servomotor dead time
T_r	s	penstock reflection time
T_{st}	s	period of mass oscillation
T_{th}	s	theoretical period of first harmonic
T_{vf}	s	minimum time of valve closure

T_{vh}	s	valve cushioning time
T_{vq}	s	valve dead time
T_{vz}	s	total valve closure time
T_w	s	water starting time
T_z	s	total servomotor closure time
t	s	time
u	–	dimensionless displacement of spool valve
u	ms^{-1}	peripheral speed of the runner ($\omega D/2$)
V	m^3	volume
x	m	distance (measured along pipe axis)
x_i	–	relative change in speed ($(n - n_o)/n_o$)
x_{ir}	–	overspeed protection setting
x_{ist}	–	dimensionless manual speed setting
x_r	mm	displacement of spool valve
Y	mm	displacement of main servomotor of turbomachine
Y_o	mm	reference value of Y
Y_h	mm	reference position of the servomotor piston, from which the motion of the piston is decelerated
Y_m	mm	maximum displacement of the main servomotor piston
y	–	dimensionless displacement of main servomotor (Y/Y_o)
β	–	runner blade angle
ΔH	m	amplitude of pressure surge
Δh	–	relative amplitude of pressure surge ($\Delta H/H_o$)
Δh_{per}	–	value of Δh corresponding to 'perfect' closure mode of wicket gates
Δp	Pa	pressure difference acting on servomotor piston
Δt	s	time increment
Δt_e	s	equivalent delay time of a governor
Δx	m	distance increment
Δz	m	elevation increment
$\epsilon_{\Delta h}$	–	perfection coefficient
ζ	–	coefficient of head loss

ρ	kg m^{-3}	mass density
σ_F	–	axial force coefficient
τ	–	torque coefficient
τ	s	half period of pipe (2L/a)
φ	–	flow coefficient
ψ	–	energy coefficient
ω	–	angular velocity

Acknowledgements

This book is based on a study of transient phenomena undertaken by Professor S. Pejovic at the request of the Association of Power Industry, Belgrade, Yugoslavia and presented in an interim report[15] (in Serbo Croation) to the Association. The Association has given permission for the publication of some of these results in this book.

The editing of this book was supported by an Academic Travel Grant provided by The British Council and The Association for International Scientific, Educational, Cultural and Technological collaboration of Serbia.

The authors are indebted to Assistant Professor A. Gajic, Faculty of Mechanical Engineering University of Belgrade, for his careful reading of the text and many useful comments, to Mr. Nebojsa Vrhovac who prepared the drawing and Mrs. J. Carrington who typed the final manuscript.

1 Introduction

1.1 General

Hydraulic transients describe the disturbances in a fluid caused during a change from one steady state to another. The principle components of the disturbances are pressure changes caused by the propagation of pressure waves throughout the hydraulic system.

In the 1960s computer based solutions of hydraulic transient problems in hydro power plants were developed. The most commonly used numerical solution, to date, is the method of characteristics which converts the two partial differential equations of motion and continuity into four total differential equations. These equations are then expressed in a finite difference form, using the method of specified time intervals, and their solution carried out on a digital computer. This method, or any other method, also requires the governing equations or performance characteristics of each component in the system in order to solve the boundary conditions.

The derivation of the basic equations of continuity and motion, together with the mathematical theory of the method of characteristics mentioned above are outside the scope of this book but are very well documented in the literature[23].

1.2 Dangers of transient regimes

Transient regimes are inevitable; every hydro power plant must at some time be started up, switched off, undergo changes of load, etc., not to mention human errors, equipment breakdowns, earthquakes, etc. It is obvious that there are many different transient situations. The engineer is only concerned with those that might endanger the safety of a plant and its personnel, cause some damage or result in operational difficulties of any kind. To set apart the cases which deserve his attention, the engineer must first clearly define what are the consequences of transient regimes he fears the most in a given hydro power plant. These might be:

(i) maximum pressures,
(ii) occurrence of local vacuum,
(iii) cavitation,
(iv) hydraulic vibration,
(v) too strong an oscillation of the water masses.

Maximum pressures during transient regimes may destroy pipelines, valves or other components, causing considerable damage and sometimes loss of human lives. In less drastic cases, strong pressure surges may cause cracks in internal lining, damage connections between steel and concrete, destroy or cause deformations to equipment (such as air valves, penstock valves or gate valves). Sometimes the damage is not noticed at the time, but results in intensified corrosion which, combined with repeated transients, may cause unexpected collapse of the penstock.

Vacuum is usually avoided, even at a high cost, because it creates stresses and strains which are very different to those produced during normal operating regimes. Vacuum may cause the collapse of thin–walled penstocks or reinforced concrete sections which were not designed to withstand such strains. This operational difficulties may create in hydro power plants and also in water cooling systems of thermal and nuclear power plants etc. Groundwater, often heavily polluted, or gases, may be sucked into the pipeline or tunnel if a vacuum phase occurs during a transient regime. This is, obviously, a great danger in public water supply systems. Dissolved gases are released from the water whenever the local pressure drops considerably and these gases may promote the corrosion of any steel sections, with consequent damage.(7)

Cavitation occurs when the local pressure is lowered to the value of the vapour pressure at the ambient temperature. All the gases within the water are released and the water starts to evaporate when the local vacuum is 96–97%. When the pressure waves reach this point, water is condensed immediately, leaving behind space filled only with very rarified gases. The water column enters this space and collides with whatever is on the side of the cavity (i.e. another mass of water or a firm boundary) resulting in a pressure surge. In this case both vacuum and strong pressure surges are present which may result in substantial damage. The main problem with regard to cavitation is that its effects cannot be estimated by the analysis with any reasonable degree of accuracy, because the various parameters are difficult to determine during the design stage.

Hydraulic vibrations, if strong enough, cause damage of pipelines, tunnels, measuring and control equipment crumble concrete and may damage any internal lining in the tunnel. If resonance occurs, the whole power plant may easily be destroyed.

Oscillations of the water masses between the storage basin and a surge tank may cause noise, concussions, suction of air, etc. Control over the plant may be temporarily lost, and if any other incident occurs at that moment, the consequences could be disastrous.

The engineer has to consider carefully all the potential dangers for his plant, estimating the weak spots. Only then should he embark upon the analysis since this usually means a considerable expense in both time and money.

The questions the engineer needs to ask are:

(i) is the safety of the plant endangered during any transient regime?
(ii) is some kind of protection required and if so, which one?
(iii) what is the most dangerous operation for the plant and is the adopted protection adequate?

1.2.1 Accidents resulting from transients

The cases summarised below are just a few of which the authors have some information.[4,5,6,7,11,13,17,18,23] It is certain that these represent but a fraction, of all accidents of a similar nature which have happened. It must be stressed that the facts regarding accidents are extremely difficult to obtain because nobody is willing to release information of that kind. This is not, of course, a desirable attitude, because only by exchanging the full account of accidents and their causes can the future disaster, hopefully, be prevented.

(i) Auto–oscillations (and subsequent resonance) caused by a ball valve have appeared, to the authors' knowledge, in some 15 different hydro power plants and pump storage schemes. Fig. 1.1 shows an air valve, installed in a surge tank chamber, which was damaged by auto–oscillation of the downstream ball valve. The oscillogram shown in Fig. 1.2 was taken during auto–oscillation of the ball valve in another power plant (the pressure recorder was installed in the surge tank chamber).

Fig. 1.1 Air valve, installed in a surge tank chamber, damage caused by auto–oscillation of a downstream ball valve

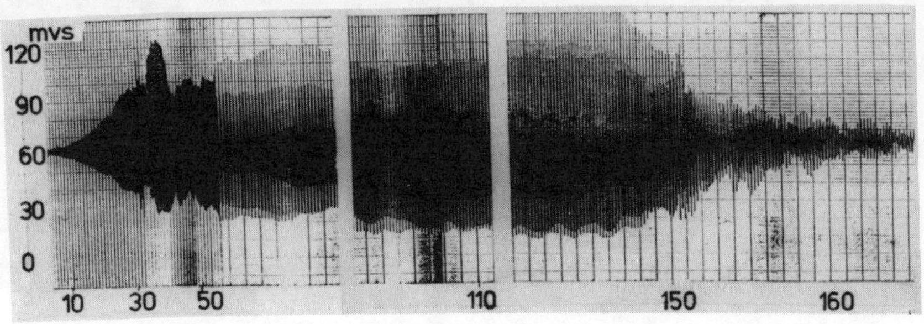

Fig. 1.2 Oscillogram showing auto–oscillation of a ball valve

(ii) Excessive oscillations of pressure and power, with resonance, have caused operational difficulties in some seven power plants.

(iii) Resonance, subsequent to a normal shut down, has occurred in three power plants.

(iv) Hydraulic vibrations and resonance due to an air valve malfunction resulted in damage to the tunnel in one power plant.

(v) Auto–oscillations due to check valve closure have caused resonance in two installations.

(vi) Reverse waterhammer in the draft tube has resulted in considerable damage in some seven hydro power plants. An example of a broken blade is shown in Fig. 1.3 and parts of the same blade, found below the runner, are shown in Fig. 1.4.

(vii) Too rapid closing–down of a butterfly valve has caused damage in four power plants.

(viii) Water column separation and the subsequent waterhammer were the cause of an accident in three cases. Figs. 1.5 and 1.6 show the consequences in one well–known case(4).

(ix) Full waterhammer pressure waves have damaged at least three major hydraulic plants.

(x) Malfunction of the relief valve was the cause of an accident in one power plant.

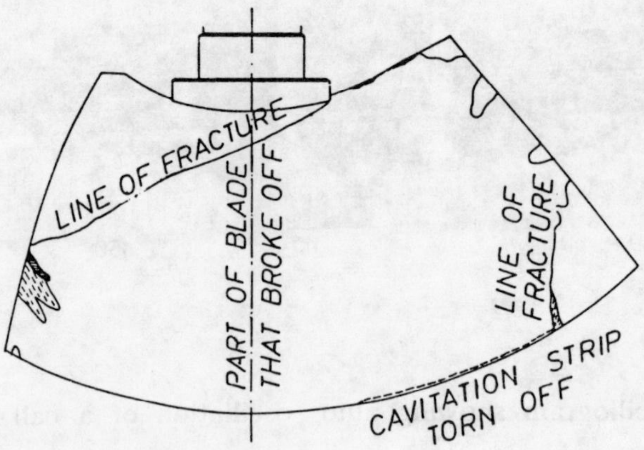

Fig. 1.3 Broken runner blade of a turbine as a result of reverse waterhammer

Fig. 1.4 Parts of broken runner blade of a turbine as a result of reverse waterhammer

Fig. 1.5 Accident caused by water column separation and subsequent waterhammer

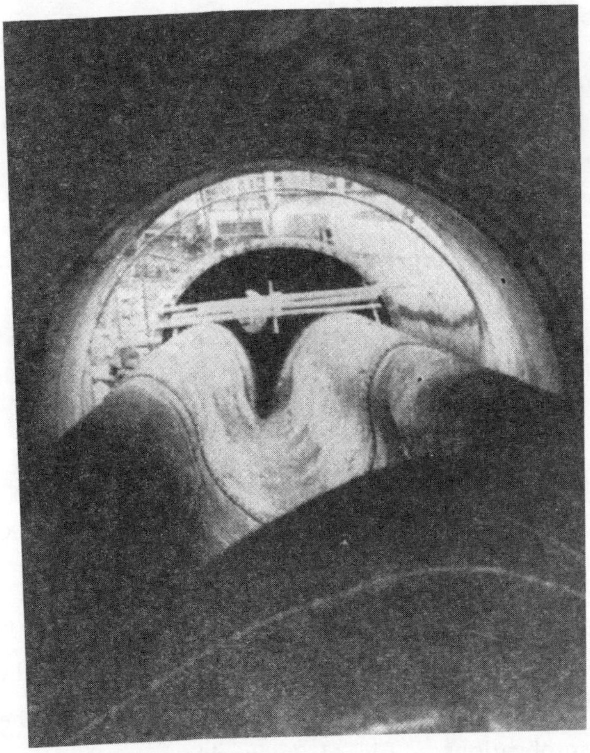

Fig. 1.6 Accident caused by water column separation and subsequent waterhammer

(xi) Transient regimes and unsuitable closure of the wicket gates were responsible for the violation of the guaranteed speed and pressure rise for one installation.

(xii) A serious breakdown occurred in the cooling system of a nucler power plant due to waterhammer.

(xiii) Too rapid closure of the wicket gates of a turbine resulted in an accident in a hydro power plant.

1.3 Principles of protection

If properly controlled, pressure surges (waterhammer) are harmless. There are many systems and devices designed specifically for that purpose. The basic principles are:

(i) to convert energy of motion into potential energy in devices such as surge tanks, reservoirs, or air vessels,

(ii) to slow down the changes by increased moment of inertia of the rotating parts, or regulation of the opening or closure of valves,

(iii) to add or subtract a limited quantity of water by installing pressure relief valves or one-way reservoirs,

(iv) to admit air through suitable valves or aeration pipes,

(v) to limit the damage by providing calibrated membranes which burst at a predetermined value of overpressure,

(vi) to change the natural frequency of a system.

The designer has to evaluate the relative merits and shortcomings of all the protection devices he may select. Usually the best solution is a combination of several devices (e.g. a power plant usually has a surge tank, but the inertia of a generator is often increased above its natural value and the timing of the wicket gates closure is adjusted to produce the best result). In his deliberations the designer must regard the safety of the power plant as his main concern, but he must also be aware of the financial implications. Fig. 1.7 provides general facts to assist the designer.

The designer should be aware of the fact that a surge tank, or an air vessel, or any change in the layout of the whole power plant changes the natural frequencies of the system. If these new frequencies are closer to the frequencies of excitations the prospects for resonance are greater. Whether the plant is safer or put in a more dangerous situation, should be settled by a careful analysis.

DEVICE	WHEN TO USE	PROTECTS AGAINST			OPERATING RELIABILITY	SPECIAL DEMANDS	PROBLEMS IN RESTARTING	FREQUENCY OF APPLICATION	COST
		HIGH PRESSURE	LOW PRESSURE	VIBRATIONS					
SURGE TANK	HIGH HEAD WITH LONG TUNNELS/ PENSTOCKS	YES	YES	YES/NO*	VERY GOOD	NONE	NONE	VERY OFTEN	VERY HIGH
AIR VESSEL	HIGH HEAD WITH LONG TUNNELS/ PENSTOCKS	YES	YES	YES/NO*	GOOD	AIR COMPRESSOR	ROUTINE CONTROL	RARELY	NOT VERY HIGH
INCREASED INERTIA (MD^2)	ALL TYPES	YES	YES	NO	EXCELLENT	NONE	NONE	OFTEN	NOT VERY HIGH
REGULATED OPENING AND CLOSURE	ALL TYPES	YES	YES	NO	GOOD	HIGH PRESSURE OIL INSTALLATION	ROUTINE CONTROL	VERY OFTEN	LOW
PRESSURE RELIEF VALVES	MEDIUM/ SMALL PLANTS	YES	NO	NO	POOR	FREQUENT MAINTENANCE AND CONTROL	ROUTINE CONTROL	NOT VERY OFTEN	LOW
ONE WAY RESERVOIR	VERY LONG TUNNEL/ PENSTOCK	NO	YES	NO	POOR	FREQUENT MAINTENANCE AND CONTROL	ROUTINE CONTROL	RARELY	NOT VERY HIGH
AIR VALVES	PROTECTION OF PENSTOCKS	NO	YES	NO	POOR	FREQUENT MAINTENANCE AND CONTROL	ROUTINE CONTROL	VERY OFTEN	LOW
AERATION PIPES	PROTECTION TUNNEL/ PENSTOCK	NO	YES	NO	GOOD	NONE	NONE	OFTEN	LOW
RUPTURE MEMBRANE	AS A LAST MEASURE OF PROTECTION	YES	NO	NO	GOOD	ARRANGE OUTLET FOR WATER DISCHARGE	REPLACE MEMBRANE	RARELY	VERY LOW

* Not an absolute protection; should be applied only after a detailed analysis

Fig. 1.7 General characteristics of protection devices

2 Hydro power plants

2.1 Elements of hydro power plants

The layouts of hydro power plants have to be adjusted to suit the local conditions. For instance the relative positions of the dam, water intake, power house, and the exit from the tailrace are very important factors in the design of a new plant. Therefore various layouts are encountered in practice. In order to identify the elements which have some relevance to the analysis of transient regimes it is sufficient to present a few typical installations.

Fig. 2.1 shows the typical layout of a high head pumped storage project. The main elements in this layout are:

1. Upper storage basin,
2. Intake/outlet structure,
3. Tunnel,
4. Aeration pipe,
5. Surge tank on the tunnel,
6. Tunnel valves chamber,
7. Air release valves,
8. Penstocks (in this case two),
9. Power house,
10. Lower surge tank,
11. Tailrace,
12. Aeration pipe,
13. Outlet/inlet structure,
14. Lower pool.

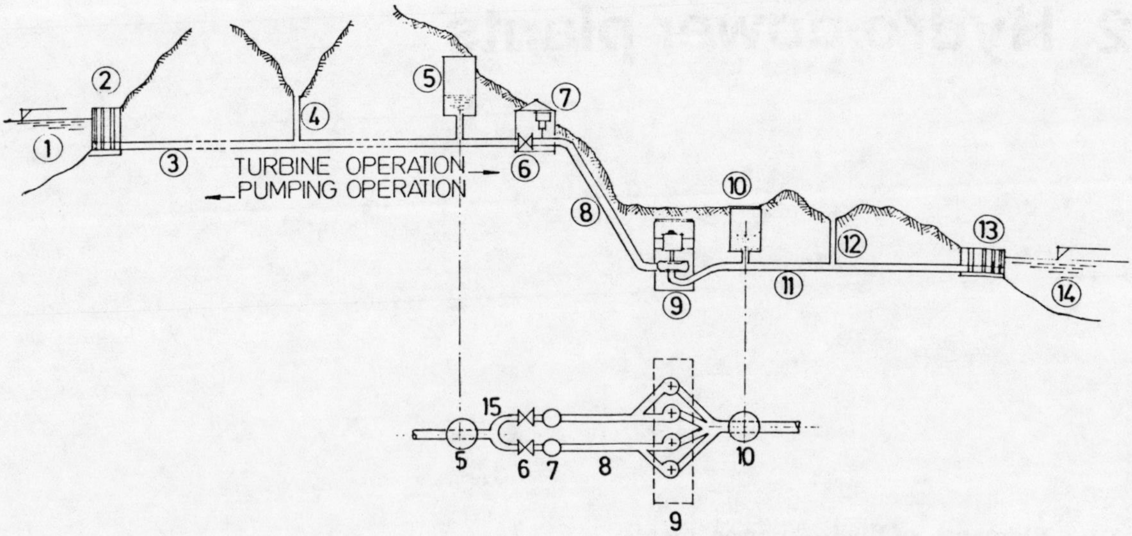

Fig. 2.1 Typical layout of a high head pumped storage project

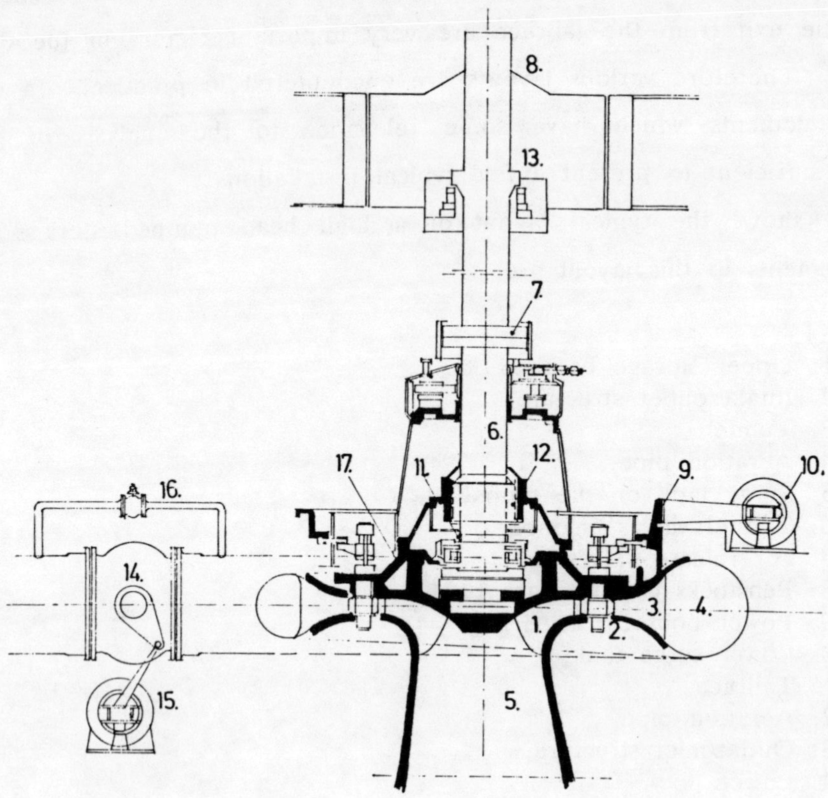

Fig. 2.2 Cross section of a reversible pump turbine

In this particular installation there are four reversible pump turbines in the power house and Fig. 2.2 shows a cross section through one of the machines. The main features are:

1. Pump turbine impeller,
2. Wicket gates,
3. Stay ring,
4. Spiral casing,
5. Draft tube,
6. Shaft,
7. Coupling,
8. Synchronous motor/generator,
9. Operating apparatus of wicket gates,
10. Main oil servomotor,
11. Labirinth seals,
12. Radial thrust bearings,
13. Axial thrust bearings,
14. Ball valve,
15. Oil servomotor,
16. Bypass pipe,
17. Turbine cover.

Many high and medium head power plants equipped with Francis type turbines are similar to the layouts shown in Figs. 2.1 and 2.2, although, in individual cases, some of the elements listed above are omitted.

Hydro power plants with Pelton type turbines usually consist of elements numbers 1 to 9 shown in Fig. 2.1, whilst the layout in the power house is shown in Fig. 2.3. The main elements of a Pelton installation are:

1. Penstock,
2. Ball valve,
3. Oil servomotor,
4. Bypass pipe,
5. Bifurcation,
6. Needle controlled by servomotor,
7. Nozzle,
8. Water jets,
9. Blades of the runner,
10. Runner,
11. Oil servomotor,
12. Turbine cover,
13. Lower pool,
14. Flow deflector.

Low head power plants usually have a considerably simpler layout. Fig. 2.4 shows such a layout for a plant with a high specific speed Francis turbine. The

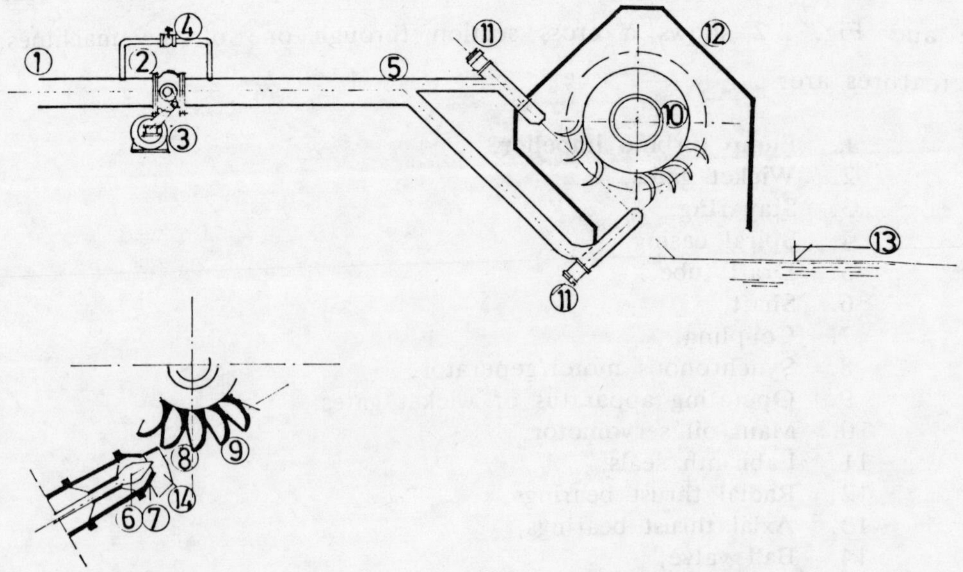

Fig. 2.3 Typical layout of a power plant with Pelton turbines

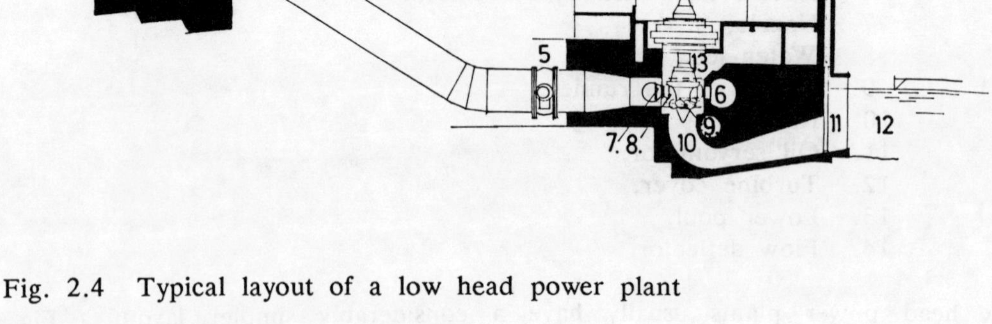

Fig. 2.4 Typical layout of a low head power plant

main features of the layout are:

1. Storage basin,
2. Entrance,
3. Intake structure,
4. Low head penstock,
5. Ball or butterfly valve,
6. Spiral casing,
7. Stay ring,
8. Wicket gates and associated apparatus,
9. Runner,
10. Draft tube,
11. Outlet structure,
12. Lower pool,
13. Shaft,
14. Generator.

A hydro power plant containing a bulb turbine consists of similar elements, arranged in a different way, as shown in Fig. 2.5.

1. Storage basin,
2. Entrance,
3. Generator,
4. Stay ring,
5. Wicket gates,
6. Runner,
7. Draft tube,
8. Lower pool,
9. Apparatus of wicket gates,
10. Power house.

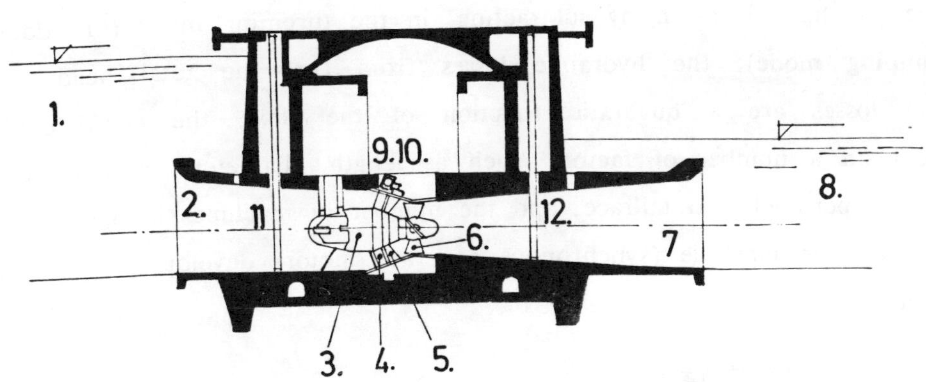

Fig. 2.5 Typical layout of a power plant with a bulb turbine

2.2 Steady state regimes

A hydro power plant is in a steady state regime if all the boundary conditions remain unchanged with respect to time. A noticeable change in any boundary condition will cause transient flow in the system which attenuates with time until eventually another steady state regime is obtained. Hydraulic vibrations (small periodic disturbances) are present in both steady and unsteady flow regimes: therefore, strictly speaking, there are no steady state regimes. However, in many practical cases hydraulic oscillations may be omitted from transient computations (but not forgotten) and the initial operating conditions of the system are assumed to be a steady state regime.

The initial operating conditions of any hydro power plant are of paramount importance for any subsequent analysis of transient phenomena. They are established by reference to the operating range of the plant and are shown by the turbine performance characteristic diagram. The total head is defined as the difference in the water elevations in the storage basin and tailrace pool. The elevation of the water surface in the storage basin may vary considerably with time, according to the current hydrological conditions and the use of the accumulated water. The water elevation in the tailrace pool depends upon the local flow into the lower basin together with the water elevation in the basin. If any factors (accumulation of ice during a particularly cold winter) can produce changes to the water elevations outside the normal ranges, these changes must be taken into account regardless of the improbability of their occurrence.

The net head is given by subtracting, in the turbining mode (or adding, for the pumping mode), the hydraulic losses from (to) the total head. These hydraulic losses are a quadratic function of the flow, the coefficient being dependent on a number of factors, such as length, diameter, friction factor for the tunnels, penstock and tailrace, and the number of machines in operation.

The electric machine (synchronous generator/motor) develops a power output P_g given by:

$$P_g = \eta_g . P \tag{2.1}$$

where η_g is the efficiency of the generator

P is the power delivered by the turbine at the generator coupling.

In the pumping mode of operation for pumped storage installations, the electrical input P_m to the motor is given by:

$$P_m = \frac{P}{\eta_m} \tag{2.2}$$

where η_m is the efficiency of the electro motor

P is the input power to the pump at the coupling.

The active power of the generator or motor can increase if the phase angle φ increases, because of

$$P_{ga} = P_g \cos \varphi \tag{2.3}$$
$$P_{ma} = P_m \cos \varphi$$

The analysis of transient regimes has to be performed for the range of values of steady state head and discharge, not just for the most frequent ones. The characteristic steady state regimes are

A. maximum available head and maximum power,
B. maximum power and maximum discharge.

Hydraulic vibrations are usually more dangerous in partial load operations. Moreover, both natural frequencies and forced oscillation of pressure or flow (such as influence of vortex core under the turbine runner) do vary with load, which makes the analysis more complex. For these reasons the analysis of hydraulic vibrations has to be performed for the complete operating range including conditions corresponding to zero efficiency.

The analysis of steady state regimes for pumped storage installations is doubled; the operating range in the pumping mode has to be determined in addition to the turbining mode. Characteristic regimes important as initial conditions are

A. maximum discharge and minimum pumping head,
B. maximum pumping head.

Again, the analysis of hydraulic vibrations has to be performed for these and various other steady state regimes.

3 Parameters relating to transient regimes

3.1 Water starting time (T_w)

This parameter is a measure of the inertia of the mass of water contained in the whole hydraulic system. It is equal to the time required to accelerate the flow from zero to the rated flow (Q_o) under the rated head (H_o). For a simple system, as shown in Fig. 3.1, with only one unit, T_w is defined by:

$$T_w = \frac{Q_o}{gH_o} \sum_{i=1}^{i=n} \frac{L(i)}{A(i)} \tag{3.1}$$

where L(i) = length of ith pipe section (m)

A(i) = area of ith pipe section (m^2)

n = number of pipe sections including spiral casing and draft tube.

For more complex hydraulic systems the value of T_w has to be calculated for each unit. For example $T_w(4)$ for the fourth turbine in the layout shown schematically in Figure 3.2 is given by

$$T_w(4) = \frac{Q_o(1)}{gH_o} \sum_{1}^{n(1)} \frac{L(i)}{A(i)} + \frac{Q_o(2.2)}{gH_o} \sum_{1}^{n(2.2)} \frac{L(i)}{A(i)} \tag{3.2}$$

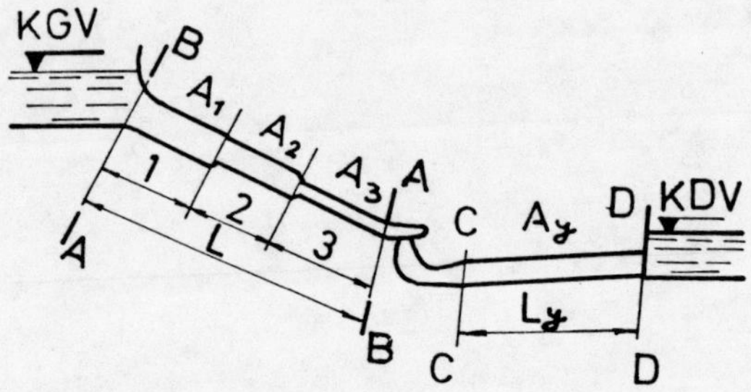

Fig. 3.1 Simple hydraulic system with one unit

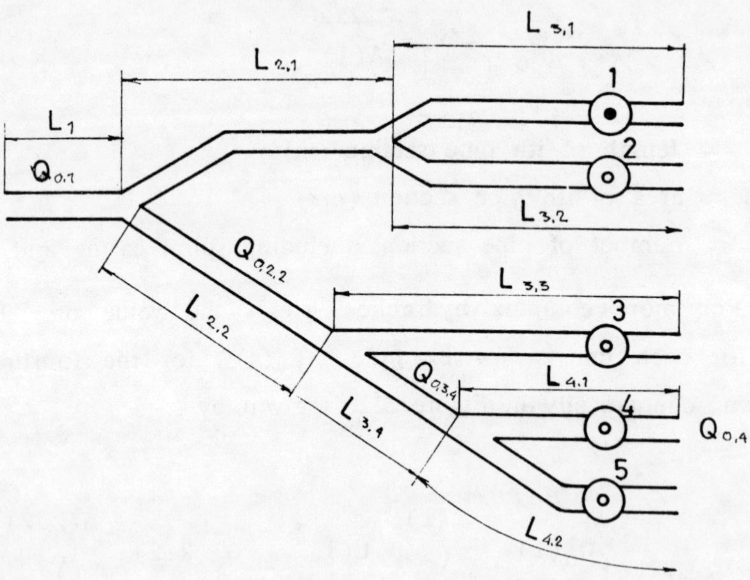

Fig. 3.2 Schematic representation of a complex hydraulic system

$$+\frac{Q_o(3.4)}{gH_o}\sum_{1}\frac{L(i)}{A(i)}+\frac{Q_o(4.1)}{gH_o}\sum_{1}\frac{L(i)}{A(i)}$$

The value of the water starting time constant for the whole plant is given by:

$$T_w(ave) = \frac{1}{N}\sum_{1}^{N} T_w(i) \tag{3.3}$$

where N = number of units.

The simple equations given above are satisfactory for the first estimates, up to the feasibility study. For the detailed design more complex calculations are necessary, especially when the true characteristics of the machines are available.

The water starting time is an important parameter in assessing the performance of any proposed hydro power plant, especially with regard to pumped storage schemes.

3.2 Unit starting time (T_a)

The combined inertia of the rotating parts of a unit (including the runner, shaft, coupling, rotor of motor/generator) has a profound influence on the transient response of any scheme, especially pumped storage, and is used in all preliminary studies. The effect is expressed as the unit starting time and defined as:

$$T_a = \frac{I_o\,\omega_o}{M_o} = \frac{MD^2}{4}\left[\frac{\pi}{30}\right]^2\frac{n_o{}^2}{P_o} \tag{3.4}$$

where I_o = polar moment of inertia of all the rotating parts (kg m^2)

ω_o = angular velocity (s^{-1})

M_o = rated torque (Nm)

MD^2 = 4 I_o (kg m^2)

n_o = synchronous rotational speed (min^{-1})

P_o = rated power at the coupling (W)

3.3 Penstock reflection time (T_r)

This is the time required for a pressure wave to travel twice the pipeline length. In the simple case of a single pipeline of length L and constant pressure wave speed (a), the reflection time is given by

$$T_r = \frac{2L}{a} \tag{3.5}$$

3.4 Theoretical period

Neglecting friction, the period of the first harmonic (T_{th}) is twice the penstock reflection time.

$$T_{th} = 2\ T_r = \frac{4L}{a} \tag{3.6}$$

Therefore the first natural frequency (f_{th}) is given by

$$f_{th} = \frac{1}{T_{th}} = \frac{a}{4L} \tag{3.7}$$

This is the lowest natural frequency of a simple pipeline. Assuming the upstream end of the pipeline is connected to a reservoir, as shown in Fig. 3.3, and the downstream end is closed (i.e. a dead end produced by a closed valve or machine) then only odd frequencies are present within the system as shown in Fig. 3.4(a) and are given by

$$f_{th}(n) = nf_{th}, \quad n = 1,\ 3,\ 5, \ldots \tag{3.8}$$

If the downsteam end of the pipeline is open (free discharge) then even frequencies are obtained as shown in Fig. 3.4(b) and are given by

$$f_{th}(n) = nf_{th}\ , \quad n = 2,\ 4,\ 6, \ldots \tag{3.9}$$

In reality most pipelines fall somewhere between the above two extremes, with a degree of throttling at the downstream end. Both odd and even harmonics may appear depending upon the throttling, effects of friction and the pipeline geometry.

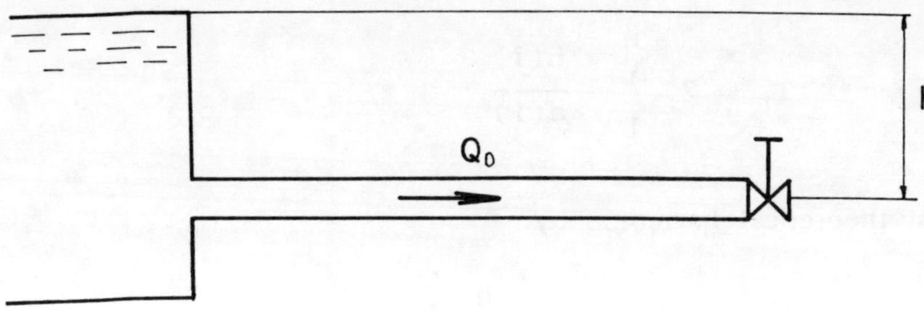

Fig. 3.3 Simple pipeline system consisting of reservoir, pipe and valve

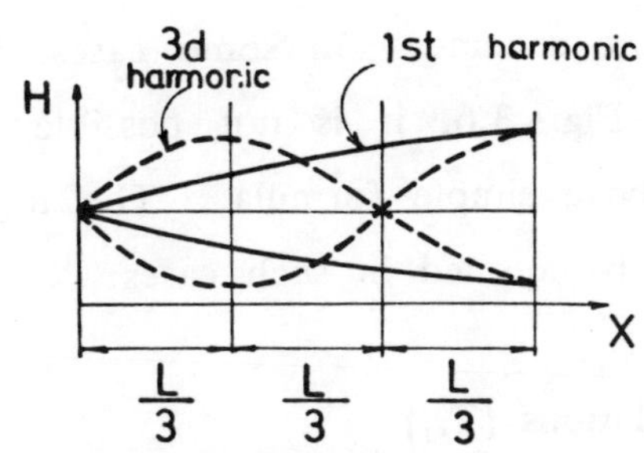

Fig. 3.4(a) Mode shapes of odd frequencies (downstream valve closed)

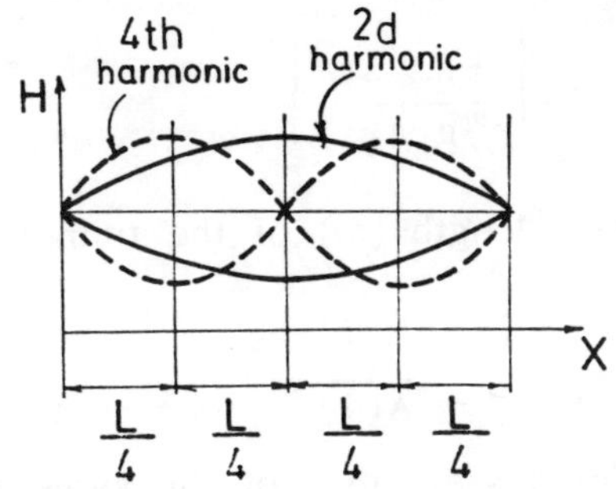

Fig. 3.4(b) Mode shapes of even frequencies (downstream valve open)

as shown in Fig. 3.5, the reflection time is given by

$$T_r = 2 \sum_{1}^{n} \frac{L(i)}{a(i)} \tag{3.10}$$

and the first theoretical harmonic by

$$T_{th} = 2\ T_r = 4 \sum_{1}^{n} \frac{L(i)}{a(i)} \tag{3.11}$$

The above equations may be used for the first evaluation of the possibility of resonance in the hydraulic system. The real natural frequencies differ from these theoretical values, so more sophisticated methods have to be applied in later stages of the design of the scheme. In some cases, where pipes form branches and loops, as shown in Fig. 3.6, it is not possible to evaluate the theoretical period by applying the above simple formula. The matrix method(5,18,23), with complex variables, should be applied in such cases.

3.5 Period of mass oscillations (T_{st})

The oscillations of the large mass of water in the low pressure tunnel between the reservoir and the surge tank in a scheme with long tunnels may be dangerous for the hydraulic system. For the purpose of the prefeasibility studies the largest period of natural frequencies (T_{st}) of such oscillations may be approximated by

$$T_{st} = 2\pi \left[\frac{L_r\ A_{st}}{g\ A_t} \right]^{0.5} \tag{3.12}$$

where L_{st} = the equivalent length (m) of the tunnel defined by

$$L_r = A_t \sum_{1}^{n} \frac{L(i)}{A(i)}$$

where A_{st} = cross-sectional area (m^2) of the surge tank

A_t = cross-sectional area (m^2) of the tunnel.

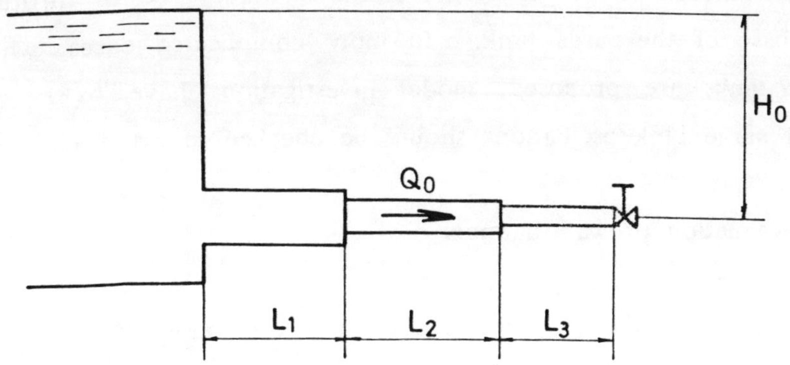

Fig. 3.5 Hydraulic system composed of a series of pipe sections

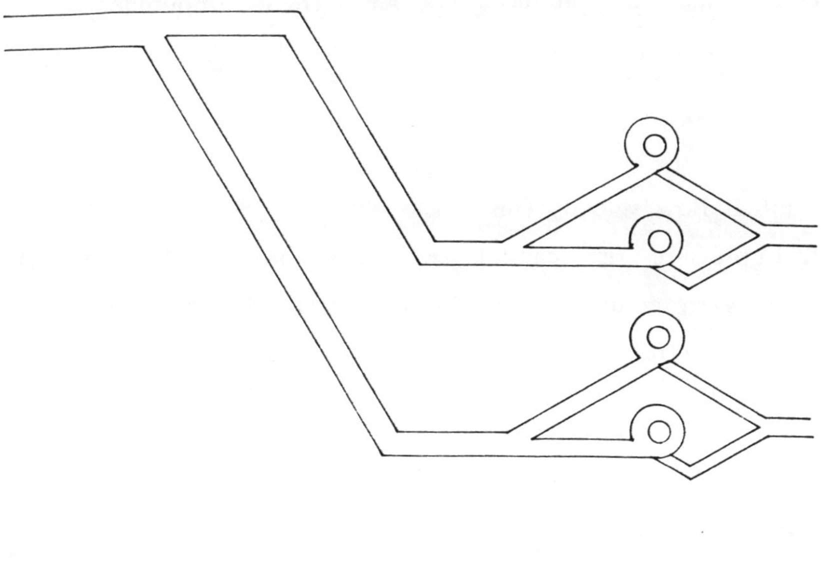

Fig. 3.6 Hydraulic system containing branches and loops

The value of the parameter T_{st} is used in designing the overall dimensions of the tunnel and the surge tank whilst avoiding the dangerous resonance range. Later as the design develops more sophisticated methods should be used which take into account the shape of the surge tank and the degree of throttling in the orifice at the base of the surge tank. In more complicated cases, when unusual types of surge tanks are proposed, model investigations have to be performed. The stability of surge tank oscillations should be checked in every case.

3.6 Joukovsky equation of waterhammer

Sudden changes in flow velocity will cause a pressure surge of amplitude, ΔH, given by the following equation usually associated with Joukovsky[16].

$$\Delta H = \frac{a}{g} \Delta c \qquad (3.13)$$

where

a = pressure wave velocity (ms^{-1})

g = gravitational acceleration (ms^{-2})

Δc = sudden change in flow velocity (ms^{-1})

The rapid closure of a valve or the guide vanes of a turbine, in a time shorter than the reflection time T_r, produces a shock wave of amplitude

$$\Delta H_s = \frac{a}{g} c_o \qquad (3.14)$$

where c_o = initial flow velocity (ms^{-1}).

The actual pressure rise cannot be greater that ΔH_s (providing that continuum of the water column is preserved) and, to find a better approximation, considerable time and effort are required.

4 Control system

4.1 Role of a governor

Transient regimes are, by definition, generated during any change of turbine machine speed and/or power. The nature and intensity of these transients depend on the overall characteristics of the hydro power scheme in question and, more particularly, on the type and characteristics of the turbine governor. A badly tuned governor may induce serious oscillations of power and pressure in the system which are very annoying to the personnel and potentially dangerous for the plant. Therefore the analysis of the transient regimes is not complete if the governor and its properties are not taken into account.

Besides the governor, there are a few other devices whose function is to react in abnormal cases in order to prevent damage. Typical examples are the valves in the surge tank chamber which are automatically closed when the water velocity in the penstock exceeds a given value. In more usual circumstances, such devices remain inactive.

Normal starting up or closing down of hydro power units are done either manually or more recently by incorporating automatic devices connected to the governor.[8,10] In modern pumped storage schemes all starts and stops in pump operating mode are executed under control of the governor. Micro–processors incorporated in the system will further enlarge the responsibilities of a governor.

A detailed description of the governor and its characteristics is included in

Appendix I.

4.2 Wicket gate stroking

The two operations which are of fundamental importance in the analysis of transient regimes are rapid closure and opening–up of the wicket gates.

4.2.1 Rapid closing down

The rapid closure of the wicket gates is necessary after a power failure in order to prevent the turbomachine's speed increasing to the runaway value. This is achieved through rapid movement of the servomotor piston as shown in Fig. 4.1 (the governor is disconnected). Fig. 4.1 shows the relative servomotor stroke, y, over the time for a typical rapid closure operation and also demonstrates the following parameters:

T_q = servomotor dead time(s)
T_f = minimum closure time from the fully open position(s)
T_h = cushioning time(s)
T_z = total closure time(s)

The servomotor piston moves with maximum speed during the time T_f until the position y_h is reached; afterwards, movement of the piston is retarded in order to attenuate pressure surges in the penstock. The value of the dead time T_q is usually 0.3 to 0.7 seconds for turbines and pump–turbines. The values of T_f , T_h and y_h have to be determined through computation of the hydraulic transients. The transition position y_h is usually between 0.1 and 0.2, but there are numerous exceptions.

The servomotor stroke–time relationship shown in Fig. 4.1 corresponds to rapid closure from full load. The same shape is also valid for quick closure from a partial load, only the starting point is moved from $y = 1$ to $y = y_o$ as shown in Fig. 4.2. The final position of the servomotor piston may not be zero and in such a case the whole procedure is followed until the prescribed value of y is reached (y_{k1} or y_{k2}).

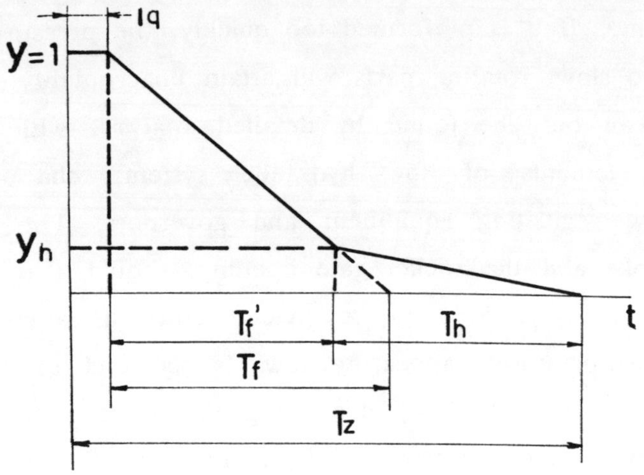

Fig. 4.1 Relative servomotor stroke against time for a typical rapid closure operation from full load

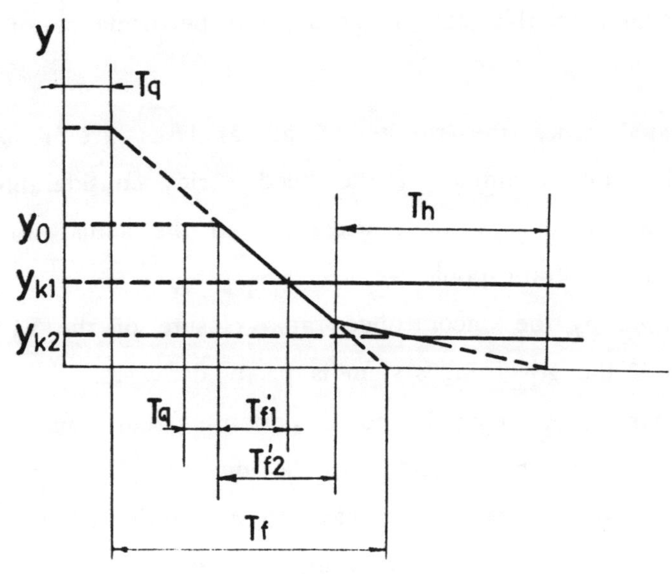

Fig. 4.2 Relative servomotor stroke against time for a typical rapid closure operation from partial load

Rapid closure is an important operation which generates high stresses and strains in the system. If it is performed too quickly, the pressure surges will be excessive; if it is to slow, rotating parts will attain unacceptably high speeds. An optimum solution can only be found by detailed analysis, which must take into account all the elements of the hydraulic system, characteristics of the turbomachines, flow–regulating equipment and governors. The relation between the servomotor stroke and the wicket gate openings should also be included. In most cases several analyses have to be made before a satisfactory solution is obtained. High overpressures, appearing towards the end of the closure time, may be attenuated or cushioned by decreasing the servomotor speed. This is described as a 'two–speed' closure. Sometimes it is necessary to use a 'three–speed' closure relationship.(9) The analysis of rapid closure has to be performed at least twice as follows:

(i) in the design phase when the manufacturer is unknown. The analysis has to be based on standard characteristics, experience and good judgement;

(ii) after bidding when the contract has been awarded. The analysis is based on detailed and reliable data regarding the performance of equipment and machine.

In more complicated cases the results of all analyses are reviewed and new solutions proposed. For example, if the head varies considerably, it might be necessary to change the closure mode according to the actual water elevation in the storage basin or the lower pool.

Slamming caused by the uncontrolled rapid closure of the wicket gates must be prevented even if the governing system is partly damaged.

The servomotor speed depends, to a certain extent, on oil viscosity and temperature, the state of the mechanism, the quality of maintenance etc. (see Appendix I). All these external factors have to be considered even in the design phase. This can be achieved in the following manner.

(i) maximum overpressure should be computed for the <u>reduced closure time</u>, T_f^H, of 10%:

$$T_f^H = 0.9\ T_f \qquad (4.1)$$

(ii) maximum overspeed should be computed for the extended closure time, T_f^{ω}, of 10%:

$$T_f^{\omega} = 1.1\ T_f \tag{4.2}$$

The shape of the closure relationship shown in Fig. 4.1 remains similar with a different time scale defined by either equation 4.1 or 4.2.[9]

4.2.2 Opening–up

If the opening–up of the wicket gates is performed too rapidly, serious vacuum conditions may be created endangering the plant. A typical procedure is shown in Fig. 4.3 which indicates the following parameters:

T_q = servomotor dead time(s)

T_{fo} = minimum opening time(s) from closed (y = 0) to full open (y = 1)

The value of the dead time, T_q, is the same as for rapid closure, between 0.3 and 0.7 seconds. The value of T_f has to be determined through computation, the usual criterion being that no vacuum is permitted even for a very short period of time.

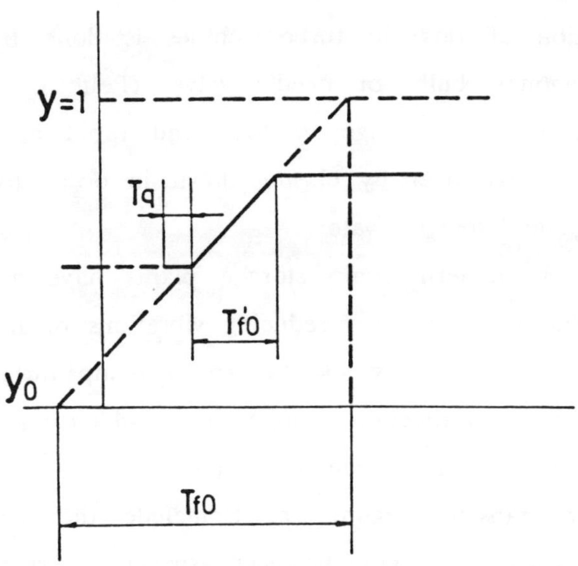

Fig. 4.3 Relative servomotor stroke against time for a typical opening–up of the wicket gates

4.2.3 *Special considerations*

The accepted modes of rapid closure and opening of the wicket gates have to be carefully determined and realised in the actual power plant. Any deviation of the accepted maximum servomotor speed might induce either excessively high overpressures or overspeed and eventually cause damage.

Maximum displacement of the piston in the spool value is adjusted so that the maximum speed of the servomotor is equal to the accepted value of rapid closure. Allowances for pressure surge effects should be given. This is done during the maintenance time, when the machines are out of operation. Afterwards, during normal operation, both rapid closure and opening of the wicket gates should be checked for the whole operating range. This should be repeated after every major maintenance or repair undertaken on the units.

Hydro power plants are sometimes installed in groups with a regulating system which takes care of all the units in a given group. Commands to the individual units are given according to the actual wicket gates opening or power developed by the unit.

4.3 Penstock valve stroking

The primary regulation of flow in turbomachines is done by the wicket gates (Francis, Kaplan, diagonal, bulb) or needle valves (Pelton). These devices are responsible for rapid closure, change of load and rapid opening. If they fail, penstock valves should take over by closing down in order to prevent the plant from eventual damage and loss of water.

Penstock valves in modern pump storage plants have a greater role; they sometimes assist in the rapid closure, reducing vibrations of the wicket gates and decreasing pressure surges. In low specific–speed pump turbines schemes, these valves act as an additional protection against surges which may be caused by the 's–shaped' characteristics of such machines in the runaway zone[14,19].

Analyses of the transient regimes must include the determination of the operation of these valves in all circumstances especially during the rapid closure regime. Although the valve may vary in type (ball, butterfly, Johnson) each consists of a flow controlling element moved by a servomotor. Typical valve servomotor stroking during rapid closure is shown in Fig. 4.4 indicating the following parameters:

T_{vq} = valve servomotor dead time(s)

T_{vf} = minimum valve closing time from the fully open position (y_z = 1) (s)

T_{vh} = cushioning time(s)

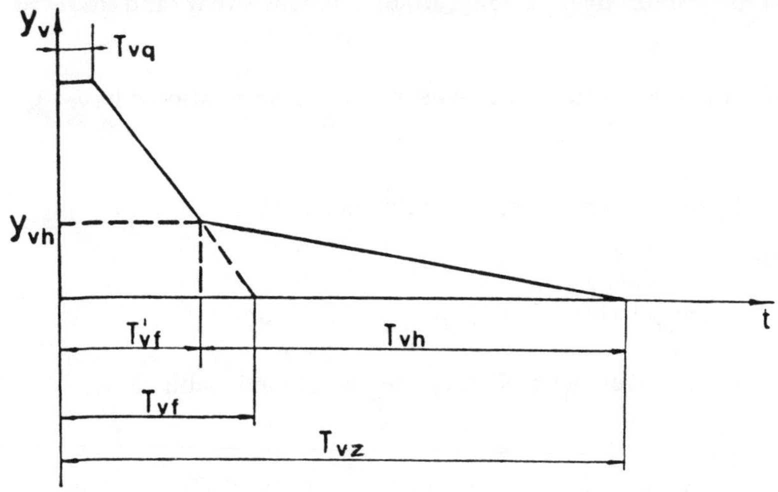

Fig. 4.4 Relative penstock valve servomotor stroke against time for a typical rapid closure

The servomotor piston moves with maximum speed during T_{vf} seconds, until it reaches a specified position, y_{vh} . The dead time is the same as for the wicket gates (0.3 to 0.7 seconds) if it is the only flow–regulating device. If there are wicket gates or nozzles then the value of T_{vq} may be considerably longer. In some cases the servomotor cannot start before the wicket gates are completely closed.

Other parameters T_{vh} , and y_{vh} should be determined on the basis of analyses, trying to obtain an optimum balance between maximum overpressure in the penstock and maximum overspeed of the turbomachine. All the elements of the plant (storage basin, tunnel, surge tanks, penstock, machine, tailrace) should be taken into account as well as the machine characteristics and the closure relationship of the wicket gates. Ideally, stresses in the system should be as uniform as possible. A 'two–speed' valve closure; helps in that direction, relieving the system in the final stage of closure; see Fig. 4.4.

Movements of a servomotor are transferred to the flow regulating part of a valve and the relation between these two movements has to be assumed during the design stage on the basis of similar installations. After the contract is awarded the manufacturer's data should be used and the computations repeated.

Slamming of the valve should be prevented, even if the pressure in an oil installation is lost due to damage or operational error. Minor changes are possible, even probable, over a long time, such as wear and tear of equipment or different oil.

As stated for the wicket gate closure the computations have to be performed twice:

(i) maximum overpressures should be computed with closure time scaled down by 10%

$$T_{vf}^{H} = 0.9\ T_{vf} \tag{4.3}$$

(ii) maximum overspeed should be computed with closure time scaled up by 10%

$$T_{vf}^{\omega} = 1.1\ T_{vf} \tag{4.4}$$

The determination of the accepted operation of the penstock valves is very important. Uncontrolled slamming of such valves has caused many disasters. Adjustment of the servomotor speed is done during the time that the units are not in operation and checked after every maintenance since deviations from the accepted values are very dangerous and therefore strictly forbidden.

4.4 Surge tank valve

The main role of these valves is to prevent emptying of the tunnel in the event of failure of the hydraulic system, for instance if the penstock is holed. In normal operation these valves are fully open. The command for rapid closing is automatically created when flow velocity in the penstock exceeds a predetermined value. The valves are closed by their servomotors, with maximum speed. Air is introduced in the penstock downstream of these valves by means of air valves or aeration pipes. Vacuum in the penstock is then considerably reduced and collapse of the pipe prevented. This should be checked by computation.

4.5 Inlet structure valve

Sometimes, when penstock valves do not exist in the power plant scheme, the intake structure valve is the protecting device of the unit, preventing the runaway when the wicket gates fail to close. The closing mode is to be calculated so as to prevent excessive surging in the system.

Behaviour of a turbomachine is different if large quantities of air are present and this fact should be included in the analysis of runaway regimes.

6.5 Interest [illegible]

[illegible]

[illegible]

5 Data for transient computations

5.1 Turbomachine

The performance characteristics of the turbomachine are necessary for transient, hydraulic oscillation and stability analyses. During the feasibility studies and the design phase these characteristics may be derived from the available literature and data corresponding to plants of similar specific speed. As the design progresses it is usual to establish the actual performance characteristics from hydraulic model tests. The extent of these tests depends on the type of turbomachine and the power plant. The investigations are generally performed in non–cavitating conditions, cavitation being investigated only for a narrow range around the normal operating zone (except for Pelton turbines where obviously cavitation effects are not important). Naturally it would be desirable for cavitation effects to be determined for the whole range of interest relating to transient analysis, but this would be too expensive. In most cases for normal plant layouts, model tests are carried out without additional cavitation experiments. However, if the plant has a long outlet tunnel without a surge tank, and the danger of reverse waterhammer cannot be ruled out, the special investigations are necessary. All known methods for reverse waterhammer computation are based upon the knowledge of non–uniformity of the flow field under the turbine runner; such information can only be supplied by new measurements.

Since the most common representation of the performance characteristics is based on unit quantities, this method is presented in this chapter together with the corresponding dimensionless coefficients. Various other representations are used in practice but each may be related to the others and must by definition be related to the dimensionless coefficients. It is therefore essential to obtain the definition of all the parameters used to specify the performance characteristics of a specific turbomachine.

In general the model tests should cover the complete operating range of the turbomachine up to and beyond the zero efficiency curve. Special attention should be given to small values (including zero) of the wicket gate opening.

5.1.1 Fixed blade turbines

This group contains turbines of the Pelton, Francis and propeller types. The model tests should establish the performance characteristics over the entire operating range up to the zero efficiency curves, in the zone of braking up to zero discharge and a section of the reverse pumping zone. At any operating point the discharge (Q), the head (H), the angular velocity (ω) and the shaft torque (M) are evaluated and recorded. The results are presented in a variety of graphical forms according to the turbine manufacturer's customs, no standardisation being imposed or agreed up to date. The two most common representations are presented here.

(i) Unit partial load curves.

Investigations are carried out for a number of openings (a_o) of the wicket gates or nozzle. (At least seven values of a_o are recommended in order to avoid large interpolations). For a given setting of a_o, values of discharge, head, speed and torque are measured directly and the results presented in the form of unit quantities.

$$\text{Unit speed} \quad n_{11} = \frac{nD}{\sqrt{H}} \tag{5.1}$$

$$\text{Unit discharge} \quad Q_{11} = \frac{Q}{D^2\sqrt{H}} \tag{5.2}$$

$$\text{Unit torque} \quad M_{11} = \frac{M}{D^3 H} \tag{5.3}$$

where
Q = discharge (m^3s^{-1})
H = turbine net head (m)
n = angular speed (min^{-1})
M = shaft torque (Nm)
D = turbine runner diameter (m)

These unit quantities are usually displayed graphically as unit discharge against unit speed, see Fig. 5.1(a), and unit torque against unit speed, see Fig. 5.1(b), both for the range of values of wicket gate opening.

(ii) Dimensionless partial load curves.

The measured parameters of discharge, head, speed and torque can also be displayed in the form of dimensionless coefficients.

Discharge coefficient $$\varphi = \frac{Q}{(\pi/8)\ \omega D^3} \quad \text{or} \quad \frac{Q}{Au} \qquad (5.4)$$

Energy coefficient $$\psi = \frac{2gH}{(\omega D/2)^2} \quad \text{or} \quad \frac{2gH}{u^2} \qquad (5.5)$$

Torque coefficient $$\tau = \frac{M}{(\pi/64)\rho\omega^2 D^5} \quad \text{or} \quad \frac{4M}{\rho D A u^2} \qquad (5.6)$$

where A = characteristic flow area = $\pi D^2/4$ (m^2)
u = peripheral speed of the turbine runner = $\omega D/2$ (ms^{-1}).

These dimensionless coefficients are displayed graphically as shown in Fig. 5.2.

5.1.2 Movable blade turbines

This group contains turbines of the Kaplan, diagonal flow (Deriaz) and bulb types. The model tests should establish the performance characteristics over the same range as specified for the fixed blade turbines. Measurements should be taken for several values of wicket gate opening (a_o) and several values of runner blade angle (β). The measured parameters are again discharge, head, angular speed and shaft torque with the addition of axial hydraulic thrust (F_a). These results are presented in a similar manner as for the fixed blade turbines.

(i) Unit partial load curves.

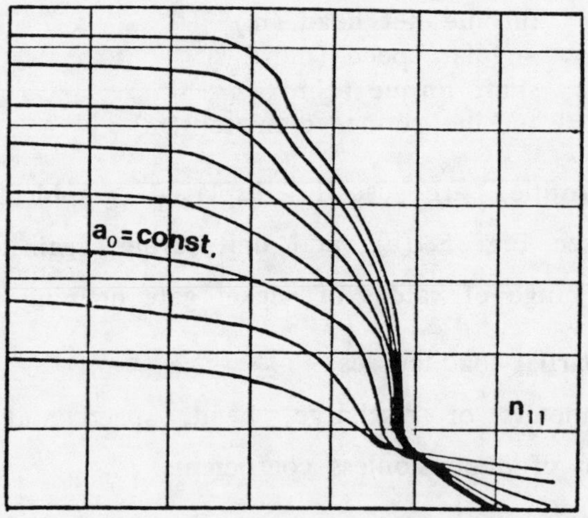

Fig. 5.1(a) Fixed blade turbine performance characteristic, unit discharge against unit speed

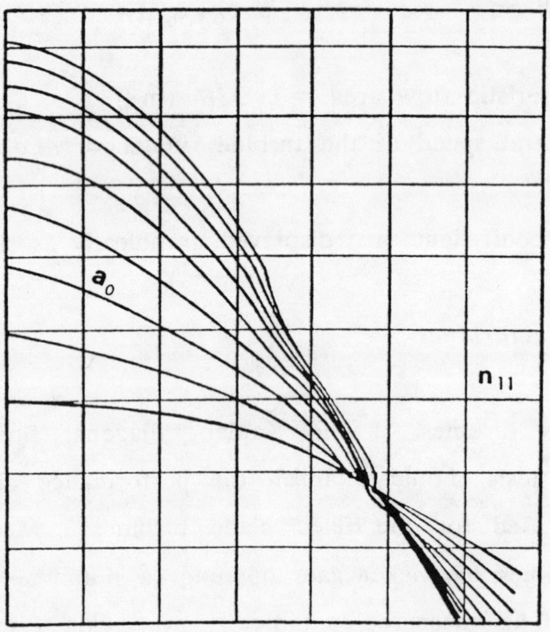

Fig. 5.1(b) Fixed blade turbine performance characteristic, unit torque against unit speed

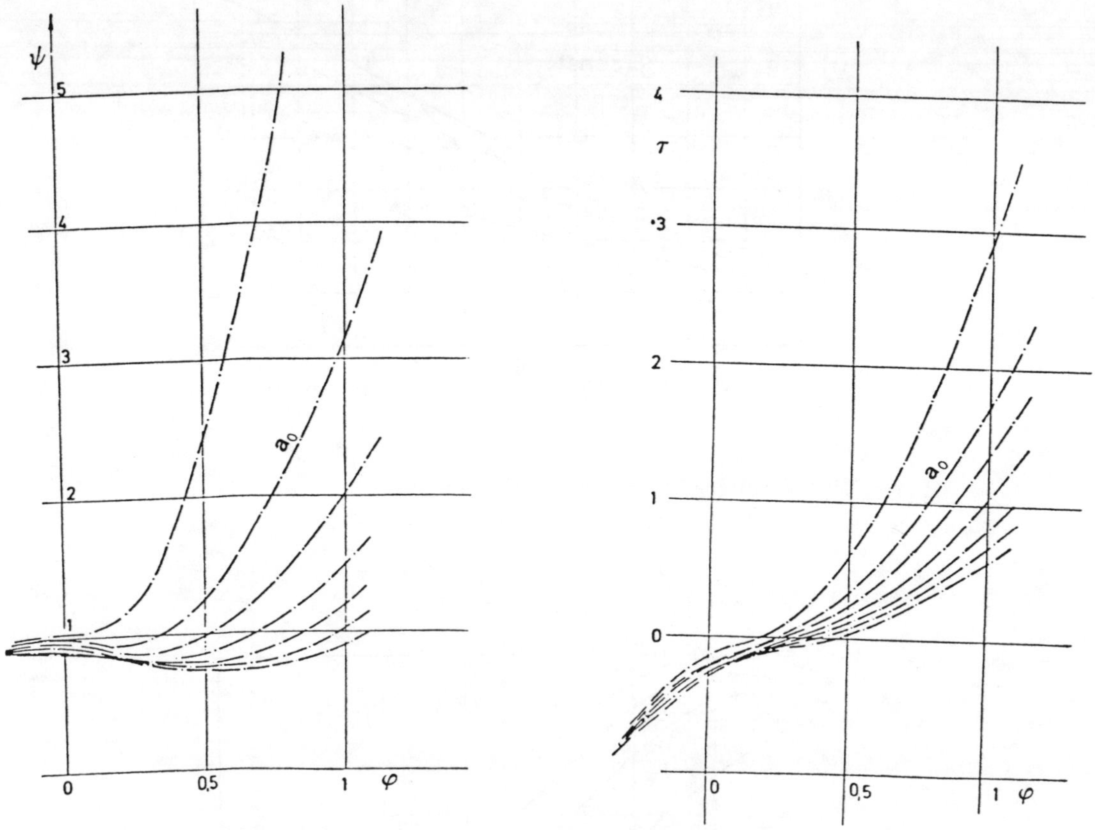

(a) dimensionless energy coefficient against dimensionless discharge coefficient

(b) dimensionless torque coefficient against dimensionless discharge coefficient

Fig. 5.2 Fixed blade turbine performance characteristic

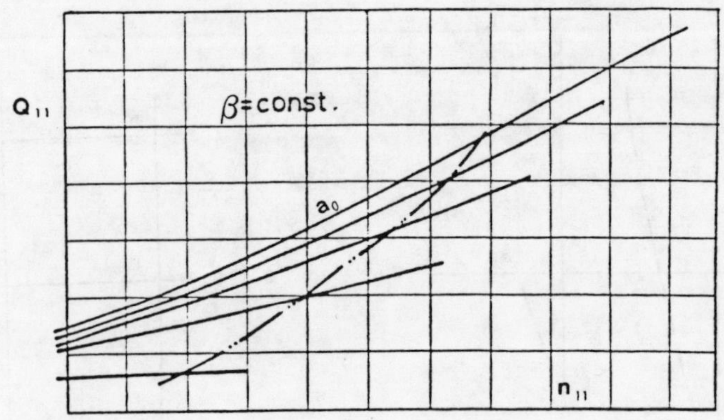

(a) Unit discharge against unit speed

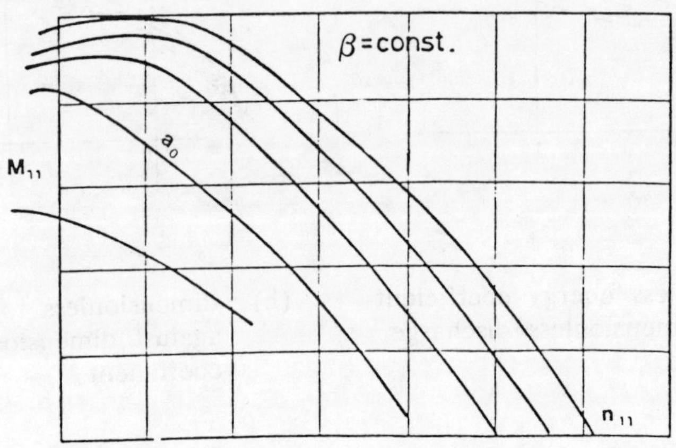

(b) Unit torque against unit speed

Fig. 5.3 Movable blade turbine performance characterisic

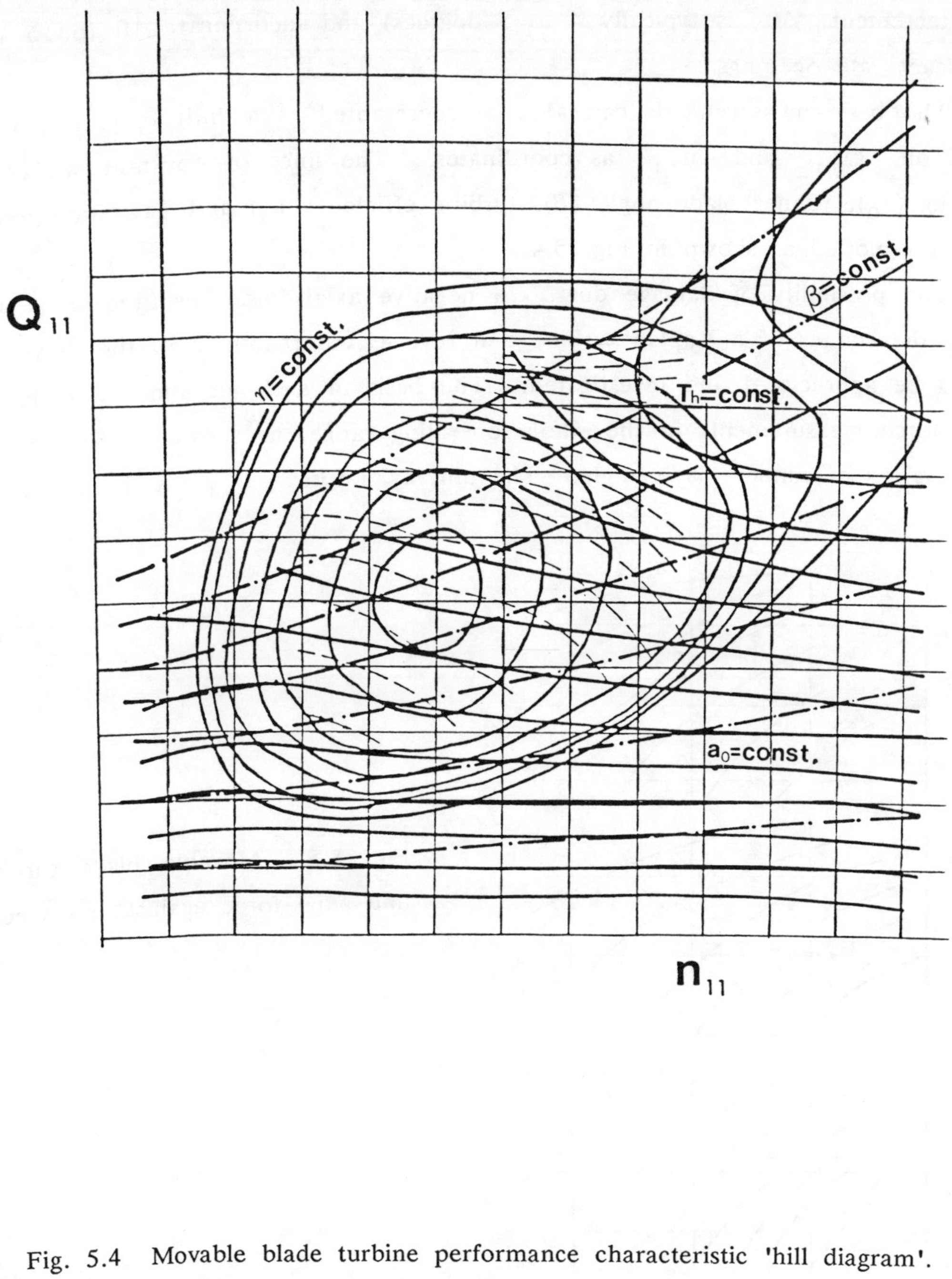

Fig. 5.4 Movable blade turbine performance characteristic 'hill diagram'.

These curves are usually represented as the so-called 'propeller curves' as shown in Fig. 5.3. For a constant value of the runner blade angle β, the unit discharge Q_{11} (Fig. 5.3(a)) and unit torque M_{11} (Fig. 5.3(b)) are plotted as functions of unit speed n_{11} and the wicket gate opening a_o. The analysis of transient regimes is possible if such curves are available for several values of β (the increment $\Delta\beta$ is typically 4 to 5 degrees) and incorporate 10 to 15 values of wicket gate openings.

The basic measurements can also be represented as a 'hill diagram', usually based on Q_{11} and n_{11} as coordinates. The lines of constant wicket gate opening (a_o), runner blade angle (β), turbine efficiency (η) and cavitation number (Th) are plotted as shown in Fig. 5.4.

The possibility of damage due to a negative axial force resulting in a thrust lifting the runner of a Kaplan turbine and reverse waterhammer in the draft tube, should be appreciated and investigated. The basis of such an analysis is data on axial force measurements in the whole operating range, as shown in Fig. 5.5 for three values of runner blade angle. The unit axial force F_{11} is defined as

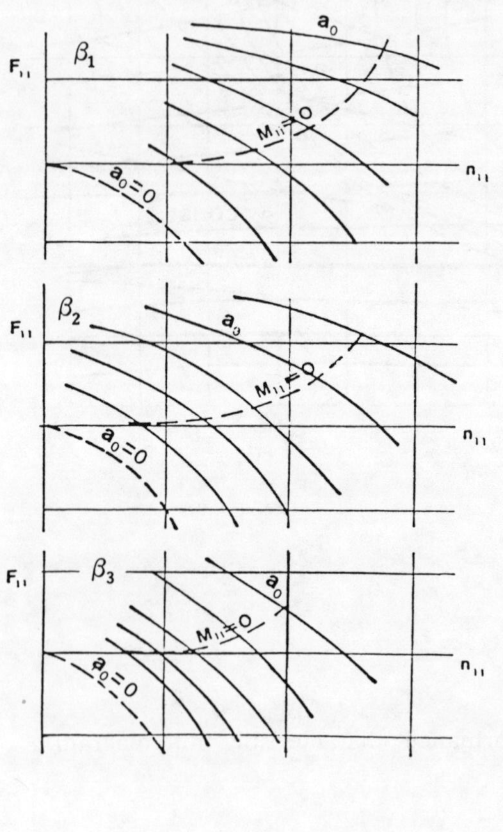

Fig. 5.5 Movable blade turbine, unit axial force against unit speed

$$\text{Unit axial force} \quad F_{11} = \frac{F_a}{D^2 H} \tag{5.7}$$

The air valves on the turbine head cover have an important influence on the analysis of reverse waterhammer because the axial hydraulic thrust could be changed considerably. However the data necessary for computations can only be determined by 'in situ' measurements.

(ii) Dimensionless partial load curves.

The 'propeller curves' can also be presented in the form of the dimensionless coefficients defined for the fixed blade turbines. These are shown in Fig. 5.6 (a) and (b) for a constant value of the runner blade angle β. As stated before several such diagrams for the range of values of β are necessary for the analysis of transient regimes.

The dimensionless coefficients can also be represented in a 'hill diagram' (Fig. 5.7) giving lines of constant wicket gate opening (a_o), runner blade angle (β), turbine efficiency (η) and cavitation number (Th).

The danger of reverse waterhammer and extensive damage due to an excessive axial thrust (all movable elements could be temporarily lifted up and suddenly dropped) cannot be appreciated without experimental data. These data are supplied by the measurement of axial force on the hydraulic scale model of the runner in all regimes which may appear during a transient state. Fig. 5.8 shows these data plotted in the form of the axial force coefficient (σ_F) against the flow coefficient (φ) for a constant value of the runner blade angle.

$$\text{Axial force coefficient} \quad \sigma_F = \frac{2F_a}{\rho A u^2} \tag{5.8}$$

As before, the axial thrust is a maximum when the wicket gates are closed, so particular attention should be given to measurements in zero flow regimes.

5.1.3 Pump turbines with fixed blades

This group consists of Francis and axial type pump turbines. As in the case of conventional turbines, measurements on small-scale models have to be carried out

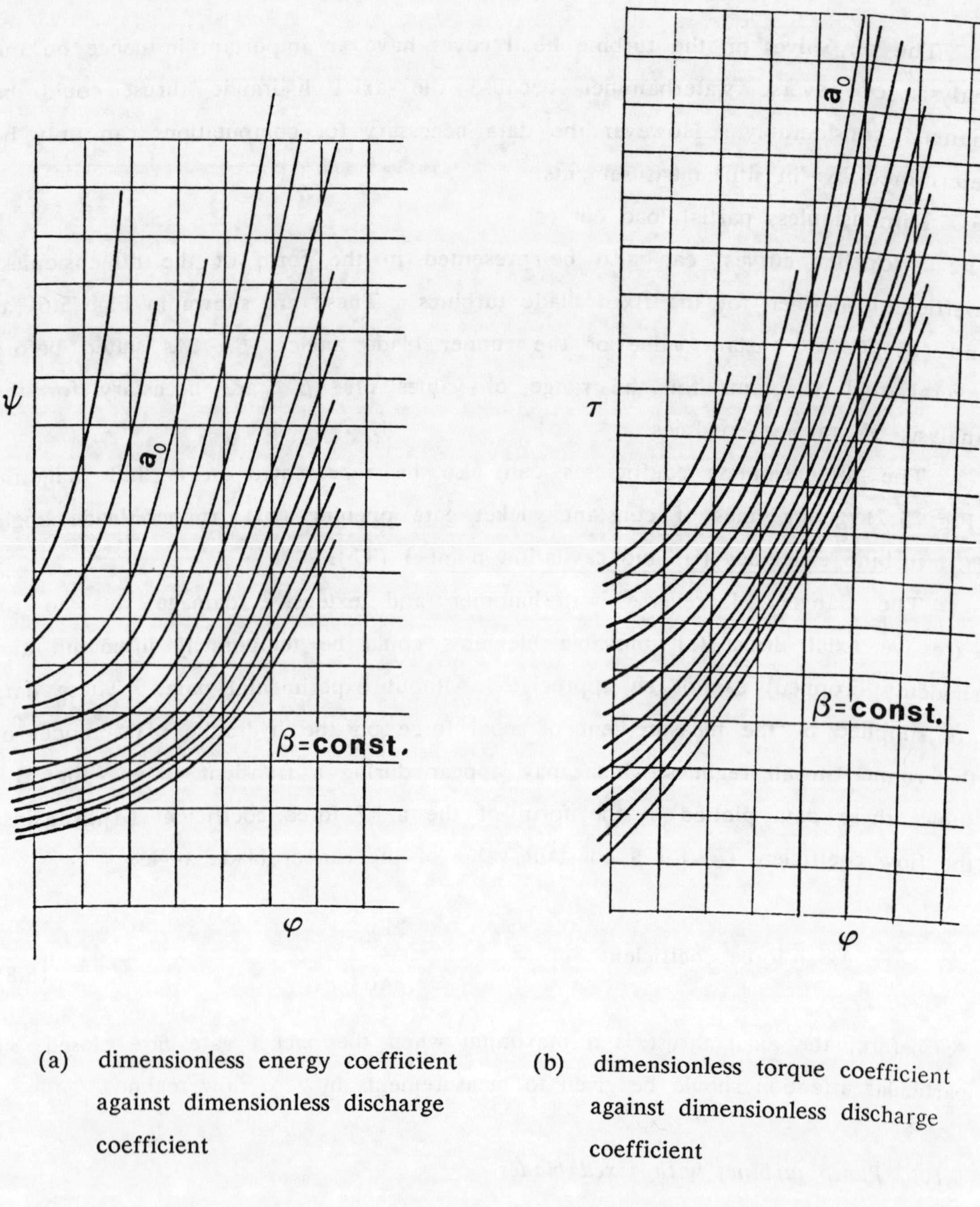

(a) dimensionless energy coefficient against dimensionless discharge coefficient

(b) dimensionless torque coefficient against dimensionless discharge coefficient

Fig. 5.6 Movable blade turbine performance characteristic

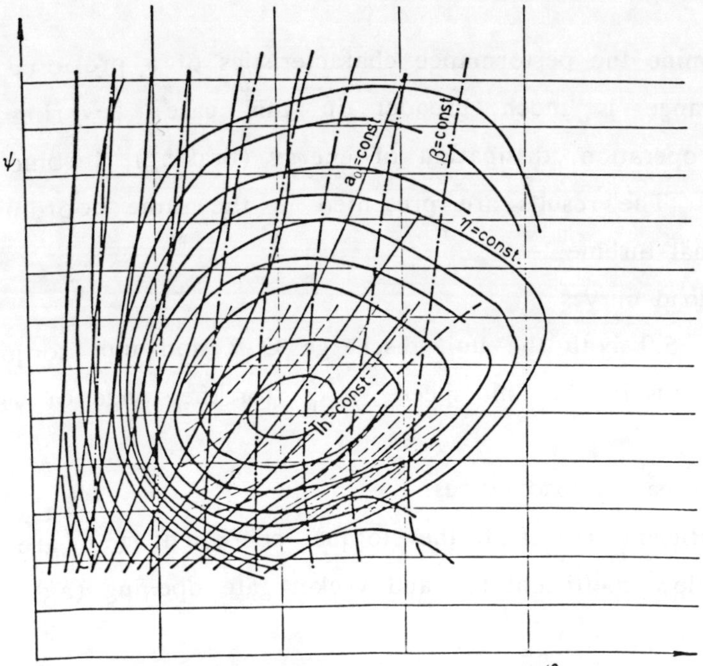

Fig. 5.7 Movable blade turbine performance characteristic, dimensionless 'hill diagram'

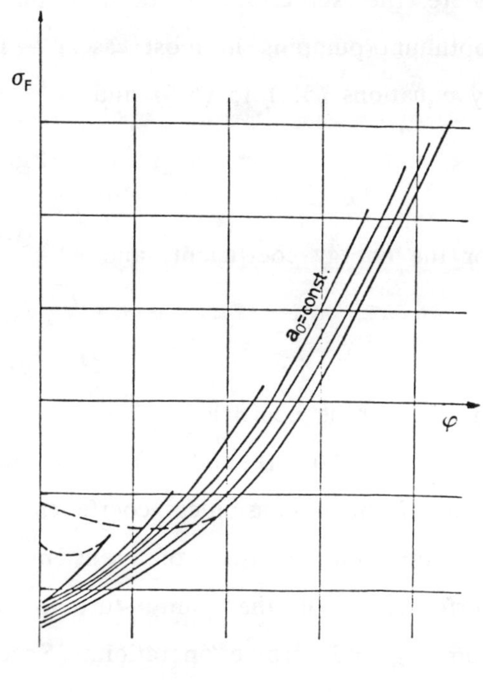

Fig. 5.8 Movable blade turbine, dimensionless axial force coefficient against dimensionless discharge coefficient

in order to determine the performance characteristics of a proposed pump turbine. The operating range is much broader in this case, covering all the four quadrants: pump operation, dissipation of energy (braking), turbine operation and reverse pumping. The results are presented in the same coordinate systems as for the conventional turbine.

(i) Unit partial load curves.

As shown in Fig. 5.9 both the unit discharge Q_{11} and unit torque M_{11} are presented as functions of the unit speed n_{11} for the range of values of wicket gate openings.

(ii) Dimensionless partial load curves.

The energy coefficient (ψ) and the torque coefficient (τ) are presented as functions of the flow coefficient (φ) and wicket gate opening (a_o), usually in the form of:

$$\frac{\psi}{\psi_o} = f\left[\frac{\varphi}{\varphi_o}, a_o\right] \text{ or } \frac{\psi_Q}{\psi_{Q_o}} = f\left[\frac{\varphi_o}{\varphi}, a_o\right] \qquad (5.9)$$

and

$$\frac{\tau}{\tau_o} = f_1\left[\frac{\varphi}{\varphi_o}, a_o\right] \text{ or } \frac{\tau_Q}{\tau_{Q_o}} = f_1\left[\frac{\varphi_o}{\varphi}, a_o\right] \qquad (5.10)$$

where the subscript 'o' defines the parameters relating to a reference regime (optimum pumping in most cases). The coefficients φ, ψ and τ are defined by equations (5.4) to (5.6) and

$$\psi_Q = \frac{2g\ HA}{Q^2} \qquad (5.11)$$

for the energy coefficient, and

$$\tau_Q = \frac{4MA}{\rho D Q^2} \qquad (5.12)$$

for the torque coefficient.

Fig. 5.10 (a) and (b) show the performance characteristics plotted in the form of the dimensionless coefficients.

For the analysis of transient regimes it is important to investigate the performance of the pump turbine when the wicket gates are closed for both pumping and turbine operation. Special care should be given to the interpolation

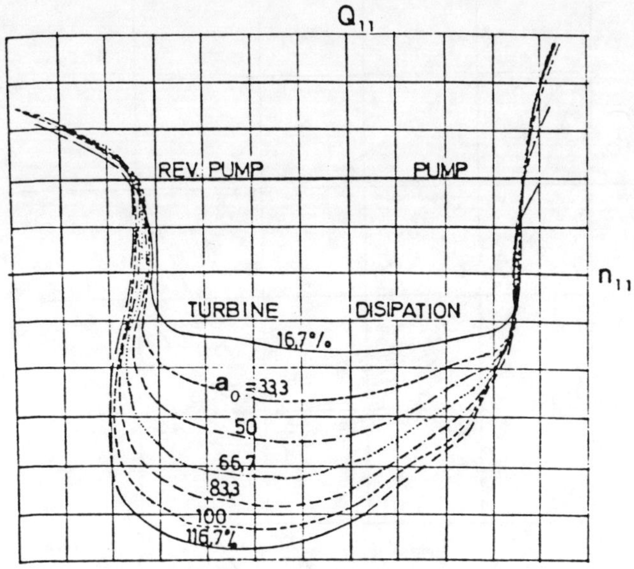

(a) unit discharge against unit speed

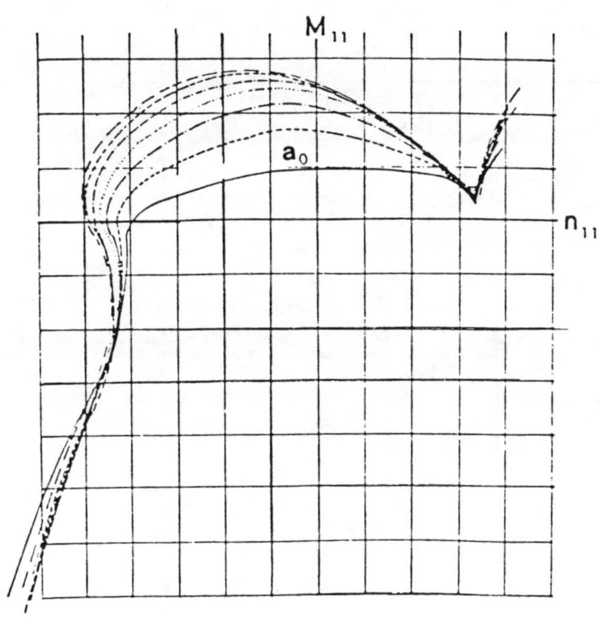

(b) unit torque against unit speed

Fig. 5.9 Fixed blade pump turbine performance characteristics

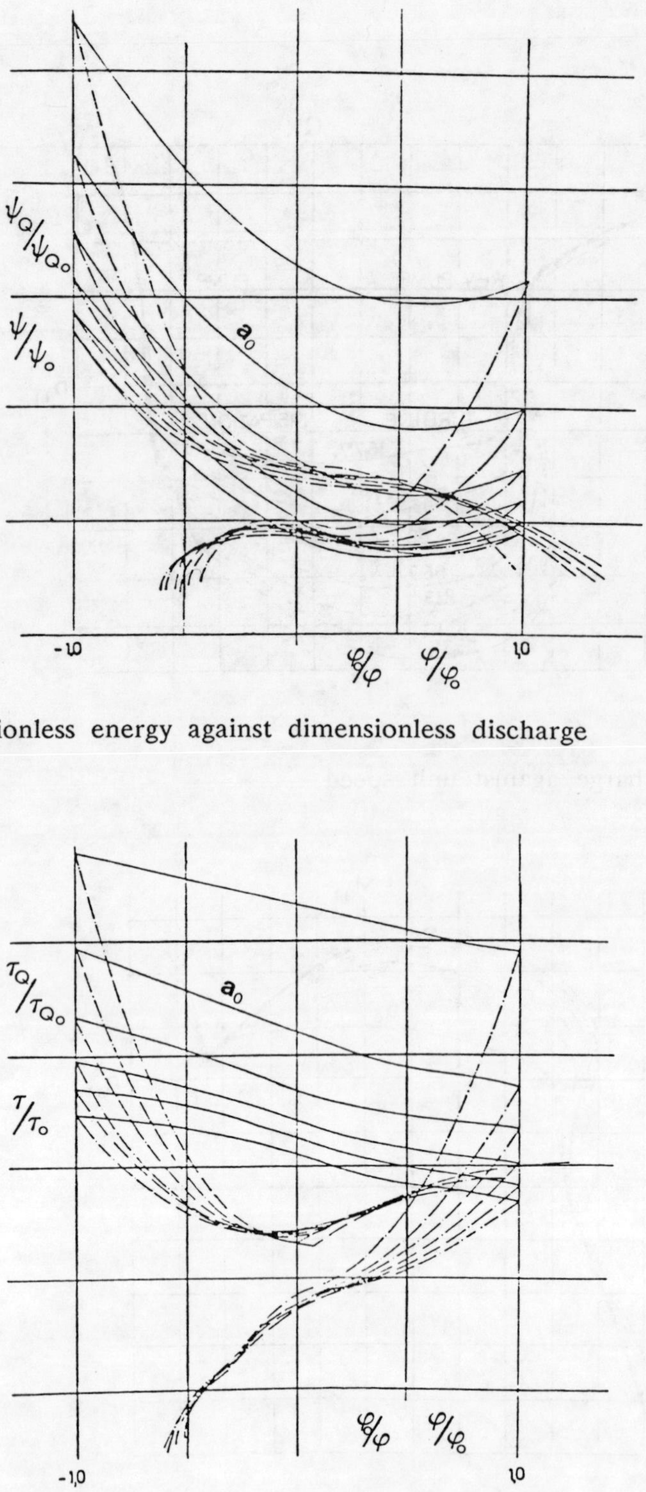

(a) dimensionless energy against dimensionless discharge

(b) dimensionless torque against dimensionless discharge

Fig. 5.10 Fixed blade pump turbine performance characteristic

between the given wicket gate curves in order to obtain an accurate representation of any intermediate wicket gate opening(3).

Again the measurements are usually carried out in non–cavitating regimes. Investigation of cavitation is, in most cases, limited to narrow ranges of normal pumping and turbining operation in order to minimise the costs. The cavitation data, if available, are very useful for the analysis of transient regimes; but even without this information the analysis may still proceed, taking into account the theoretical knowledge on cavitation phenomena.

If the tailrace (referring to turbine operation) is long, without a surge tank incorporated in it, then the possibility of reverse waterhammer occurrence should be analysed. Different methods may be used but all of them require knowledge of the unevenness of the flow field under the runner as the basis of such calculations.

5.1.4 Pump turbines with movable blades

This group contains diagonal (Deriaz), Kaplan and bulb pump turbines. As above, measurements on small–scale models are essential in order to establish the performance characteristics of the machine in all four quadrants (pumping, energy dissipation, turbining and reverse pumping) for the range of wicket gate openings and runner blade angles. The results of the model tests are displayed as above.

(i) Unit partial load curves.

The unit discharge (Q_{11}) and unit torque (M_{11}) are presented as functions of unit speed (n_{11}) for the various values of wicket gate opening (a_0) and blade angle (β). Several diagrams are usually prepared, similar to those shown in Fig. 5.9, each corresponding to a constant value of the blade angle.

(ii) Dimensionless partial load curves.

The energy coefficient (ψ) and torque coefficient (τ) are presented as functions of the discharge coefficient (φ) for various values of wicket gate opening (a_0) and blade angles (β). Diagrams are usually prepared in the form of

$$\frac{\psi}{\psi_0} = f\left[\frac{\varphi}{\varphi_0},\ a_0,\ \beta\right] \qquad (5.13)$$

and

$$\frac{\tau}{\tau_0} = f_1\left[\frac{\varphi}{\varphi_0},\ a_0,\ \beta\right]$$

or $$\frac{\psi_Q}{\psi_{Q_o}} = f_2 \left[\frac{\varphi}{\varphi_o}, a_o, \beta \right] \tag{5.14}$$

and $$\frac{\tau_Q}{\tau_{Q_o}} = f_3 \left[\frac{\varphi}{\varphi_o}, a_o, \beta \right]$$

For a given value of blade angle (β), a diagram similar to Fig. 5.10 is prepared as the 'propeller diagram'.

5.2 Valves

Experimental data on head losses in valves, measured in steady-state conditions, have to be used for transient analyses. The head loss coefficient (ζ) is given by

$$\zeta = \frac{1}{K^2} - 1 \tag{5.15}$$

where K is the flow coefficient for the valve which is usually presented graphically as a function of the valve opening (y) as shown in Fig. 5.11. The head loss, H_L, across the value is given by

$$H_L = \zeta\frac{V^2}{2g} = \zeta\frac{Q^2}{2gA^2} \tag{5.16}$$

where Q = discharge (m^3s^{-1})

A = valve area (m^2)

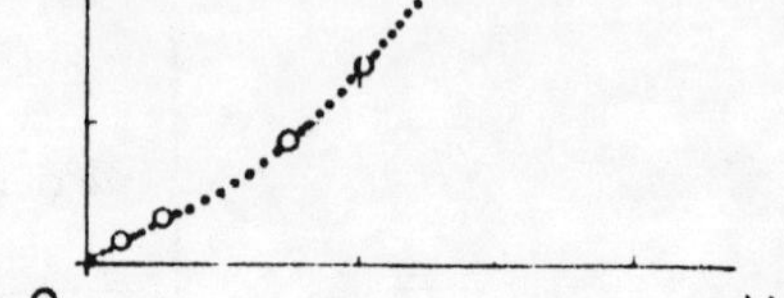

Fig. 5.11 Flow coefficient for a valve against valve opening

5.3 Wave speed

In order to calculate the value of the wave speed for the various sections of the hydraulic system it is necessary to obtain the following data:

(i) for steel penstocks : diameter, thickness, modulus of elasticity, Poisson ratio and support conditions.

(ii) for tunnels : dimensions of steel linings (as for steel penstocks) and characteristic dimensions of the concrete together with modulus of elasticity and Poisson ratio for steel, concrete, rock or earth appropriate to each tunnel.

Gas bubbles substantially decrease the value of the wave speed, so it is important to know the quantity of air (or other gases) free or dissolved in the water. When the local pressure falls below saturated pressure, as in the case of the turbine draft tube, the dissolved air will be released reducing the wave speed. The value of the wave speed may be less than 100 ms^{-1} and therefore these free gases have a great influence on waterhammer phenomena.

5.4 Hydraulic losses

Friction losses should not be ignored in any transient analysis although they are seldom of great importance. Therefore any formulae for the friction head loss derived for steady−state computations may be used, the Darcy−Weisbach equation being the most popular. The value of the friction factor is usually determined from the Colebrook−White implicit formula given in all handbooks on hydraulics.

For any properly designed and constructed hydraulic system, local head losses are of minor importance. The approximate value of the head loss coefficients are given in handbooks and the final computations should be performed using their measured values.

5.5 Surge tanks

Various designs of surge tanks are used in practice ranging from cylindrical, differential, with or without lower and upper expansion galleries to a gate shaft.

The shape is sometimes peculiar and any space that may be temporarily occupied by water during a transient operation should be regarded as a surge tank (e.g. aeration pipe, roller–gate shaft, access shaft). Any surge tank should be checked for overtopping of the water surface.

For the analysis of transient phenomena in systems incorporating a surge tank the following data relating to the surge tank are necessary:

(i) shape of the surge tank (horizontal area as a function of elevation),
(ii) location of surge tank in the hydraulic system (may be various alternatives in the feasibility study),
(iii) overflow arrangements,
(iv) connection between the surge tank and the hydraulic system.

The head loss, H_L, at the entrance to the surge tank is given by

$$H_L = K_{st} Q_{st}^2 \tag{5.17}$$

where $K_{st} = K_{in}$ for $Q_{st} > 0$
$= K_{out}$ for $Q_{st} < 0$
$Q_{st} =$ flow into surge tank (> 0)
$=$ flow out of surge tank (< 0).

The value of the coefficients K_{in} and K_{out} can be evaluated theoretically in simple cases, but it is desirable to determine the exact values by measurements on hydraulic models. In many situations the throttling of the flow in or out of the surge tank is rather strong in order to keep the dimensions within acceptable limits. Special orifices, usualy non–symmetrical, are designed to satisfy the limits and the evaluation of the head loss coefficients K_{in} and K_{out} over the whole operating range on a hydraulic model is essential. Strictly K_{in} and K_{out} are not constant, but vary with the ratio of the flow in or out of the surge tank to the flow through the plant. If available these data should be taken into account in the transient analysis.

During particularly violent surges water might be spilled out of the surge tank, intermittently. This effect has to be simulated in the analysis and the necessary data are weir width and overflow constant.

Surge tanks do not attenuate hydraulic oscillations; in fact, in some cases, they even act as a resonator. For the analysis of hydraulic oscillation and stability of the whole system it is necessary to have additional data on the surge tank in order to calculate the wave speed in the surge tank (see Section 5.3).

5.6 Relief valves

Pressure surges are frequently regulated by relief valves usually connected to the spiral casing and opened synchronously with the rapid closing of the wicket gates, thus allowing part of the flow to escape from the system thereby preventing excessive overpressures. After the wicket gates are closed the relief valves are then slowly closed down. Such an arrangement has a profound influence on the transient regimes and must therefore be simulated carefully. The required data are the same as for any valve (see Section 5.2) and supplied by manufacturers on the basis of laboratory tests.

5.7 Draft tube gate

If a draft tube gate is incorporated in the plant and located close to the machine the combined effect can change the performance characteristics of the machine. In this case, the investigations on the model must extend to incorporate the draft tube gate. The results of measurements are usually given in the form of diagrams similar to those presented in Section 5.1 with an additional parameter (y), the displacement of the draft tube gate. Partial load curves are presented for several values of draft tube gate position.

5.8 Branching

The head losses in the vicinity of branches are usually neglected in transient flow analyses since these branches are carefully designed and constructed. However, if it is considered that these losses might be significant, the head loss coefficients have to be determined through experiments in the laboratory for different flow ratios in the branches.

5.9 Governor

The governor data required for normal transient analysis are discussed in Chapter

4 and Appendix I. For stability analysis and hydraulic oscillation computation, further data are necessary.

Y_m = maximum servomotor stroke,

β_m = maximum inclination of runner blades (Kaplan, Deriaz or bulb units),

T_K = time constant of the spool valve for the runner servomotor,

a_o = f(y), relation between wicket gates opening and servomotor strokes,

φ = $f(a_o)$, on cam function.

5.10 Motor generator

The data relating to the electrical machine necessary for transient analysis are usually expressed as the polar moment of inertia of the rotating elements (MD^2).

5.11 Earthquake

The analysis of the system response to a possible earthquake should be based upon characteristic earthquake oscillations recorded in the area where the plant will be constructed. The selection of the typical oscillogram is done by experts in seismology and transient analysis.

6 Transient regimes

6.1 Classification of transient regimes

The transient regimes in hydro power plants may be divided into the following three distinct types depending on the boundary conditions[5]:

(i) transient flow in closed conduit systems,

(ii) transient flow in open channels,

(iii) combined cases.

This book only deals with the first type which are described by nonlinear partial differential equations of the first degree[5,23]. The water is regarded as a Newtonian fluid with a constant value of the modulus of compressibility. Deformations of the pipes are assumed to be small, hence completely within the elastic range, implying that Hooke's Law is valid. The partial differential equation together with the appropriate characteristic for the various boundary conditions have to be solved numerically and several methods have been proposed in the literature[23]. However, costs are rather high even when the most modern computers are used and therefore simple methods for transient analysis of the following special cases have been investigated:

(i) slow oscillations of the water masses,

(ii) hydraulic vibrations due to small harmonic disturbances of flow or pressure.

Slow oscillations of the water masses appear after a normal start−up, change of load or closing down of a plant. Pressure surges are not violent, so the deformations of the pipes are negligible and the water may be regarded as a non−compressible fluid. Any change of flow is immediately transmitted to the other end of the pipeline, the water column behaving as a rigid shaft, hence the name of the method of analysis 'rigid water column theory'. The partial differential equations are transformed into ordinary differential equations which can be solved by very efficient methods.

Using the rigid column theory the savings in computational costs are high; the difficulty is how to decide when it is appropriate to use the method. The answer is not straightforward; the problem is that the elasticity of the system is significant whenever the derivative of pressure over distance ($\partial h/\partial x$) cannot be dropped out of the equations. This term might be important even if pressure amplitudes are small, providing the frequency is high. Each case needs to be considered separately before an answer can be given. The diagram shown in Fig. 6.1[11,16] may assist in making the correct decision in simple cases. The diagram is prepared for a single pipeline with a constant head at one end and a valve at the other end. The relative value of pressure surge, $1 + \Delta h = 1 + H/H_o$, is plotted against 2μ defined as

$$2\mu = \frac{2mT_w}{T_f} \qquad (6.1)$$

where T_w = water starting time (Section 3.1)

T_f = closing or opening time of the downstream valve

$$m = \frac{T_f \cdot a}{2L}$$

$$\tau = \frac{2T_w}{T_f}$$

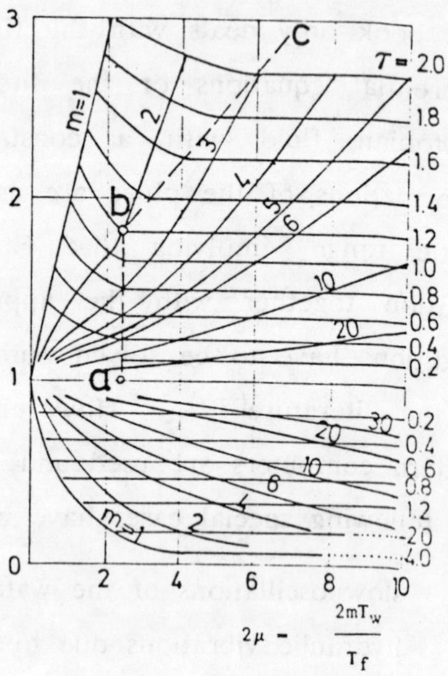

Fig. 6.1 Diagram indicating when rigid column theory may be used for a single pipeline

The discharge at the valve was varied linearly. For a constant value of τ the values of $\Delta h/h$ were calculated varying the value of m, achieved by changing the wave speed (a) until m = 1. (i.e. $T_f = 2L/a$). As a result curves of $\Delta h/h_o$ as a function of m were plotted for constant values of T_w/T_f. These diagrams were then used as the basis for Fig. 6.1.

For closing down operations (i.e. $\Delta h > 0$) the rigid water column theory may be applied to cases falling to the right of the broken line abc on Fig. 6.1. The elasticity of the system has no significant influence on pressure drop due to opening up operations ($\Delta h < 0$).

For more complex cases the possibility of applying cheaper calculation methods must be proved. The analysis should be performed using the full waterhammer equations and simplified equations. Comparison of the results will establish which method should be used. In more important cases the designer should consult the specialists in transient phenomena.

Hydraulic vibrations of a system, excited by small periodic changes of flow or pressure at a few points, have to be analysed using the linearised equations of waterhammer. By applying modern methods of analysis, the designer is able to determine frequency of free vibrations, amplitude, phase angle and damping of hydraulic vibrations. Such an analysis should be made even during the feasibility study, with assumed data. Later during the final design stages, such an analysis is essential and must be performed with great care.

Another classification of transient regimes may be made as follows:

(i) normal operating conditions, expected and executed as scheduled. These are not dangerous for the system,

(ii) emergency operating conditions, unexpected but probable. These may cause inconveniences but should not damage the system,

(iii) catastrophic cases, unexpected and highly improbable. These cases may cause damage to the system and should not be considered as a basis for the design.

6.2 Normal operating conditions

The hydro power plant should be designed to operate without difficulties during any of the following normal operating conditions (shown diagrammatically in Fig.

6.2.(1). These apply to any one unit of the plant and the effect on the remaining units should be considered.

(i) normal steady state conditions either turbining, pumping or speed no−load,

(ii) standstill to turbine operation and loading (line 1),

(iii) turbine operation to synchronous compensation (line 2) and return to turbine operation (line 3),

(iv) stopping from turbine operation (line 4),

(v) synchronous compensation to closure (line 5),

(vi) standstill to pump operation and loading (line 6),

(vii) stopping from pump operation (line 7),

(viii) pump operation to synchronous compensation (line 8) and back to pump operation (line 9),

(ix) synchronous compensation to closure (line 10),

(x) pump to turbine operation (line 11). Usually the machine will be stopped first (line 7) and then started from standstill to turbining (line 1),

(xi) power failure during turbine operation followed by rapid closure of the wicket gates (line 12),

(xiii) power failure during pump operation followed by rapid closure of the wicket gates (line 13).

The transient regimes associated with the operating sequences listed above are considered normal and should not cause any inconveniences in a properly designed and constructed hydro power plant. To achieve this, all these transient regimes must be carefully analysed during the design stages of the proposed scheme. The whole range of operation has to be covered in the simulations, but special attention should be given to the extremes, maximum and minimum net head, discharge and developed power, even if such regimes do not occur frequently. Maximum changes of load from speed no−load to full load should be analysed with maximum speed of the servomotor piston. In due course other changes have to be analysed (such as partial load changes, small variations in head, flow and power), but these situations are usually less dangerous for the system than the extreme cases.

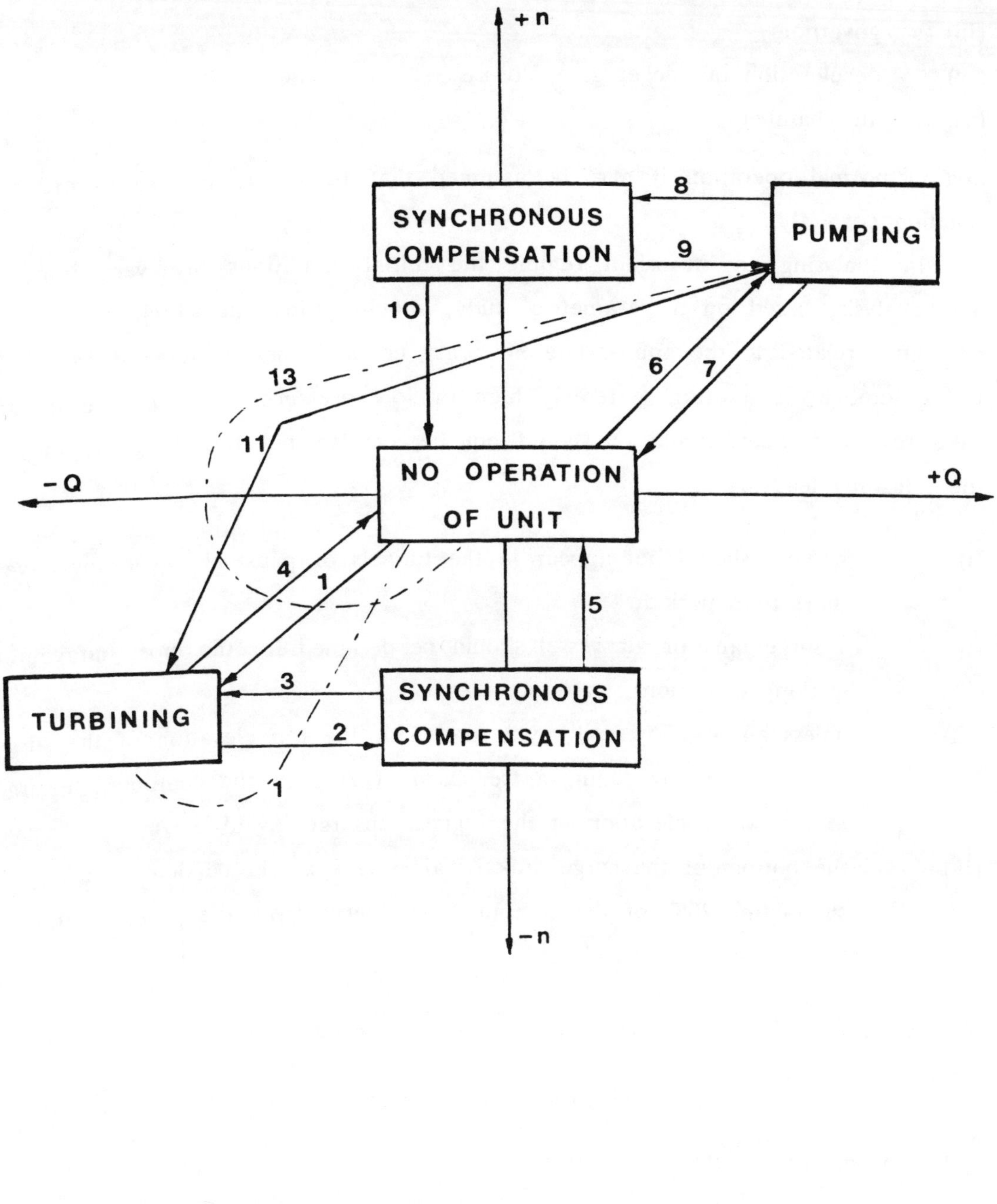

Fig. 6.2 Normal modes of operation for hydro power plants

If necessary, devices for the protection against damaging effects of pressure surges have to be selected, designed, built and later maintained. These may include one or several of the following devices:

(i) surge tanks

(ii) relief valves

(iii) governors

(iv) deceleration of wicket gate closure

(v) air chambers

and in normal operation it may be assumed that these devices are in order and function correctly.

In analysing any transient regime the initial conditions are very important. The analysis, based on a parametric study, should point out which set of initial conditions related to the appropriate boundary condition is the most dangerous for the system, by generating extremely high or low pressures. There are no strict rules regarding what should or should not be avoided but the designers may use the following leads:

(i) vacuum should not appear in the tunnels or penstocks even for a very short time period,

(ii) a surge tank or air vessel should be designed for the most unfavourable working conditions,

(iii) unless an overflow spillway is provided, the top elevation of the surge tank is determined by adding a freeboard of 20% of the computed upsurge to the water elevation at the highest upsurge level,[22]

(iv) the bottom of the surge tank or air vessel should be determined by subtracting 20% of the computed downsurge from the water level at the lowest downsurge.[22]

The various modes of normal operation are discussed in detail in the following sections.

6.2.1 Standstill to turbine operation

This process is shown schematically in Fig. 6.3. Point 1 corresponds to the initiation command and the time period denoted by T_{pr} is used for preparations. The wicket gates or nozzles start to open at point 2 and full opening, point 3, is reached in T_o seconds. When the runner reaches full synchronous speed, point 4, the opening of the wicket gates or nozzle is decreased slightly until the machine is successfully synchronised with the electrical network (point 5). The wicket gate or nozzle opening y_p corresponds to idle turning. The turbine can accept load from point 5 onwards. During the above sequence the pressure decreases in the penstock and increases in the draft tube. The time delays T_{pr} and T_{st} should be kept as low as possible.

For Kaplan turbines the runner blades are rotated to the position corresponding to idle operation; line β in Fig. 6.3.

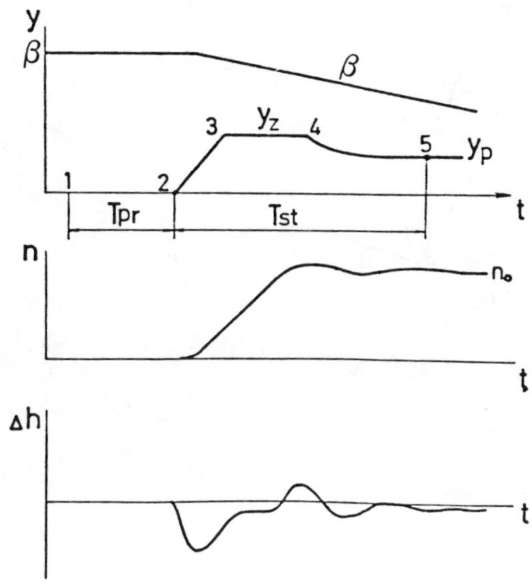

Fig. 6.3 Standstill to turbine operation

6.2.2 *Turbining to synchronous compensation operation*

This process begins by bringing the turbine to the 'speed no–load' condition. The wicket gates are then completely closed as shown by line 1–2 on Fig. 6.4, but the unit is not disconnected from the electrical network. The torque on the shaft assumes negative values because the operating regime is in the so–called 'braking zone' of the turbine characteristic. In turbines with movable blades the axial hydraulic force (F_a) becomes negative causing an upthrust and vacuum may appear between the closed wicket gates and the runner. The danger of

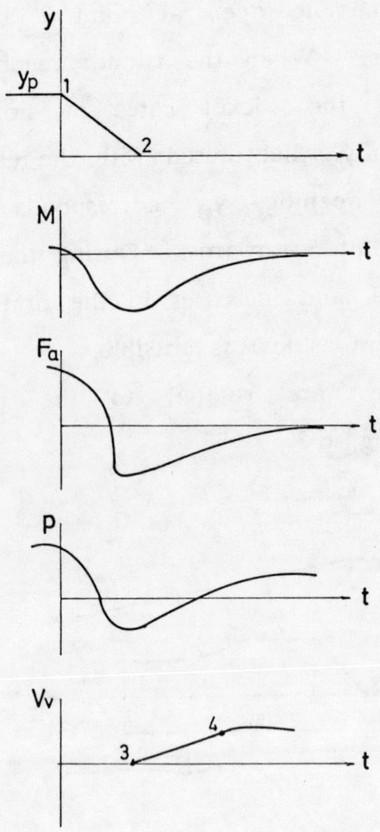

Fig. 6.4 Turbining to synchronous compensation operation

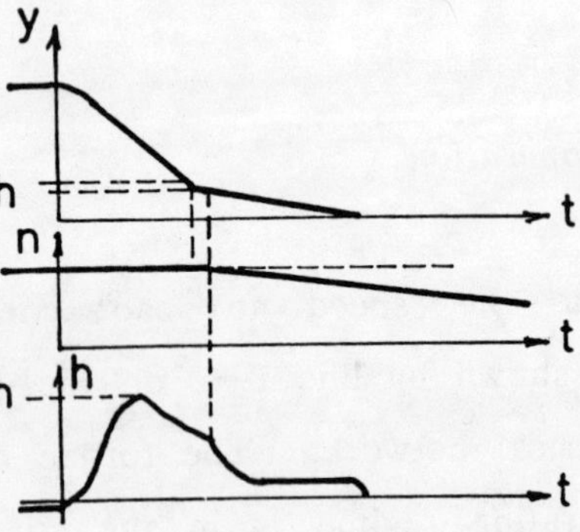

Fig. 6.5 Normal stopping from from turbining operation

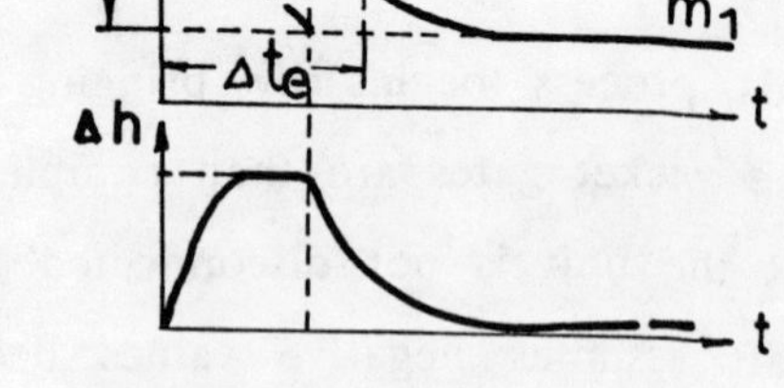

Fig. 6.6 Change of load in turbine operation

reverse waterhammer occurring is high and therefore the wicket gates should not be closed too fast. It is also important to check if the upthrust is sufficient to lift all the rotating parts (rotor, shaft and runner) which would create considerable damage.

After the wicket gates are completely closed, the space around the runner is filled with compressed air, point 3, and the water is pushed down. The axial force is rather small and the pressure depends on the tail water elevation. The volume of air (V_v) is maintained constant from point 4 onwards. In machines with movable blades the runner is completely closed during this operation The runner rotates in air in order to decrease the losses of power compared with rotation in water. For Pelton turbines such an operation is simple since no compressor is required.

6.2.3 Normal stopping from turbining operation

This operation starts either by giving a 'stop' signal to the governor or by actuating the turbine/governor protection which in turn generates the 'stop' command. The servomotor closes down the wicket gates with maximum speed as shown in Fig. 6.5. The turbine remains connected to the electrical network up to the line of zero torque (zero efficiency), then it is disconnected and the rotational speed decreases. Pressure in the penstock increases during this process whilst the pressure in the draft tube decreases.

Pelton machines are stopped in a similar manner, the movement of the needle gradually closing the flow through the nozzle.

6.2.4 Change of load in turbine operation

The governor keeps variations of rotational speed within acceptable limits during any change of load. Maximum stresses occur in the system when the spool valve is completely open and servomotor piston moves at maximum speed. The transient regime is shown in Fig. 6.6 and characterised by

(i) initial (m_o) and final (m_1) values of torque with corresponding positions (y_o) and (y_1) of the servomotor piston.

(ii) change of torque $m(t)$ between m_o and m_1 which depends on servomotor speed, head and waterhammer effects. At the beginning of the closure of the wicket gates it is possible for the torque to be rising before starting to decrease. The broken line shown in Fig. 6.6 represents the change of torque without the influence of waterhammer.

(iii) the pressure surge Δh.

All these parameters may be represented by one equivalent delay time Δt_e, defined as [9]

$$\Delta t_e = \frac{\int_0^{\infty} \left| m(t) - m_o \right| dt}{\left| m_o - m_1 \right|} \tag{6.2}$$

and shown visually on Fig. 6.6.

For small changes of load equation 6.2 may be linearised as

$$\Delta t_e = 0.5 \frac{\left| m_o - m_1 \right|}{\partial m/\partial y} T_f + \frac{m_o + 0.5\ \partial m/\partial x_i}{\partial m/\partial y} \cdot \frac{\partial q}{\partial y} T_w \tag{6.3}$$

where $\partial m/\partial y$ = change of torque with wicket gate opening (or position of servomotor piston)

$\partial m/\partial x_i$ = change of torque with change of relative rotational speed (the self regulation coefficient)

$\partial q/\partial y$ = change of discharge with servomotor stroke

and the value of the above rates of change are determined from the turbine characteristics.

The relation

$$\Delta t_e = f(m_o, |m_o - m_1|) \tag{6.4}$$

represents the universal characteristics of regulation; see Fig. 6.7 which presents the curves for various values of m_o. The value of Δt_e can easily be found for a given value of m_1. In the case of small changes of torque, equation 6.3 is more appropriate.

The universal characteristic is usually computed for the rated head. As Δt_e depends on actual head, it is advisable to prepare such diagrams for maximum and minimum heads also if the range is large.

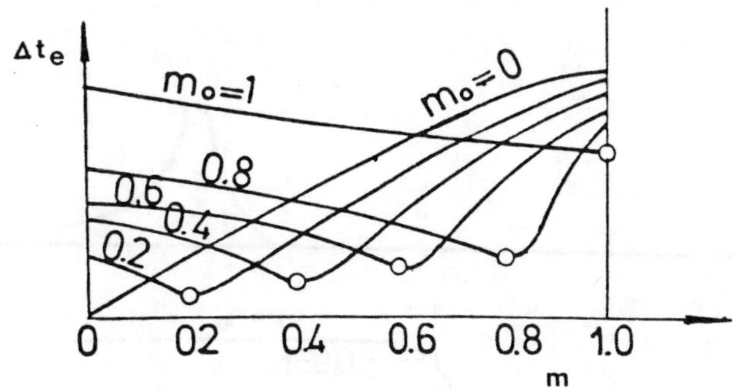

Fig. 6.7 Change of load in turbine operation; universal characteristic of regulation showing variation of the equivalent delay time as a function of the torque

6.2.5 Standstill to pump operation

This operation is fully automatised in large modern pump storage plants, as shown in Fig. 6.8. The water is first depressed by injecting compressed air, then the runner is set in rotation and its speed synchronised to the electrical network (phase 1). In the second phase the air is let out, the water level rises and the runner is filled with water. The wicket gates are opened in phase 3, the water flows into the penstock provoking a temporary pressure rise in the penstock and a corresponding pressure drop in the draft tube. For an analysis of transient effects it is only necessary to consider phase 3.

If the standstill to pump operation is performed with the help of another hydraulic machine (a pony turbine or another turbine) this part of the whole process should be analysed as an ordinary turbine start.

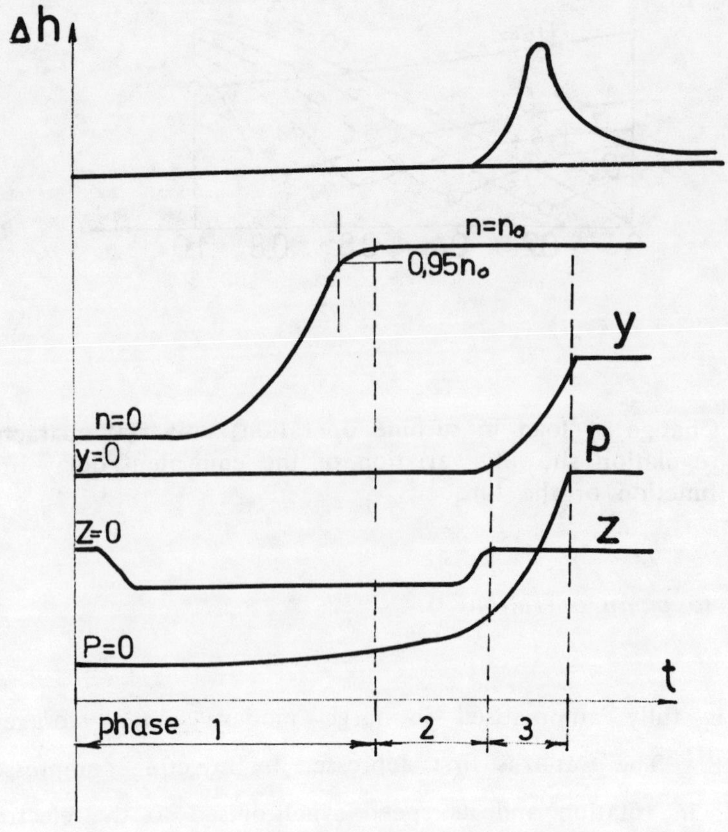

Fig. 6.8 Standstill to pump operation

6.2.6 *Normal stopping from pump operation*

This operation is usually performed by closing down the wicket gates at a constant rate. If the amplitude of the pressure surges are too high towards the end of the closure, then the movement of the piston should be decelerated as

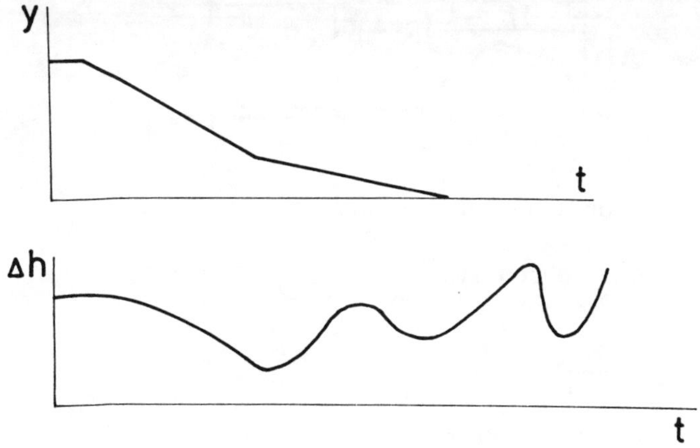

Fig. 6.9 Normal stopping from pump operation

shown in Fig. 6.9. Maximum and minimum pressures have to be determined by computation. The same procedure is valid in the case when the machine is stopped by closing down a penstock valve.

The optimum closure mode is found from computations by varying the closure rates and times.

6.2.7 *Load rejection from turbine operation*

The unit may suddenly reject load either during start–up (unsuccessful start) or during steady state operation due to some disturbances. In this case the wicket gates or nozzle close completely in T_f seconds as shown in Fig. 6.10. Both the rotational speed and penstock pressure head increase to maximum values during the wicket gate closure. T_q is the dead time of the governor. When the rotational speed drops back near to the synchronous speed, the wicket gates are opened to the position corresponding to idle rotation (y_p) and the unit is ready for synchronisation.

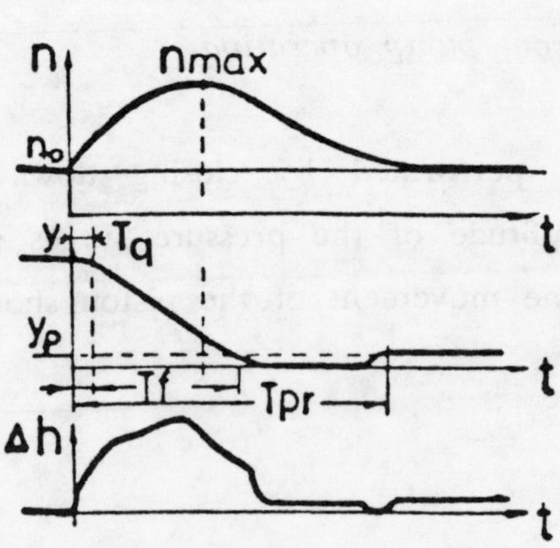

Fig. 6.10 Load rejection from turbine operation

Maximum values of temporary speed rise

$$x_{i\,max} = \frac{n_{max} - n_o}{n_o} \tag{6.5}$$

and pressure surge

$$\Delta h_{max} = \frac{\Delta H_{max}}{H_o} \tag{6.6}$$

should be computed for the worst case, which is usually load rejection of all the units of the plant simultaneously. If the values of $x_{i\,max}$ and/or Δh_{max} are unacceptable or the total time of this operation (T_{pr}) is too long, the speed of the wicket gate or nozzle closure should be changed accordingly. If this is not successful (because the penstock is too long, for instance) then more drastic changes have to be introduced, such as increasing the inertia of the generator rotor or the installation of a synchronous relief valve.

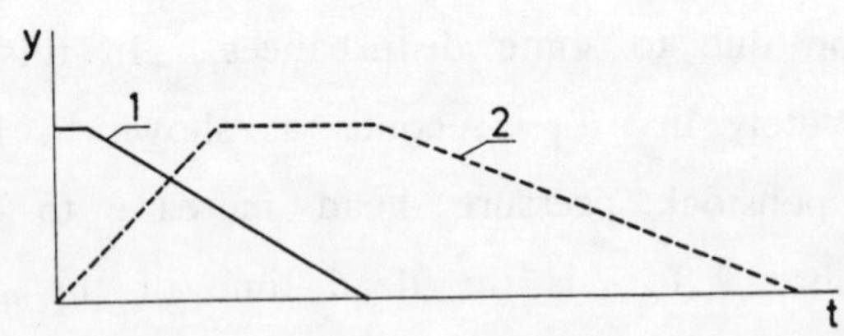

Fig. 6.11 Load rejection from turbine operation; operation of a relief valve

The operation of a relief valve is shown in Fig. 6.11. When the turbine rejects load the valve starts to open-up and part of the flow is diverted through the valve relieving the turbine. The wicket gates may be closed rapidly without causing excessive pressure surges. Later in the sequence the relief valve is closed slowly. Maximum valves of x_i and Δh have to be obtained by computation and, by appropriate operation of the relief valve and wicket gates, both may be kept as low as necessary.

Such a relief valve is usually incorporated in hydro power plants equipped with Francis type turbines. In those with Pelton machines, a flow deflection device prevents excessive rotational speeds and excessive pressure surges can then be avoided by appropriate closure of the nozzles.

6.2.8 *Turbine quick closure*

This case is demonstrated in Fig. 6.12 and is similar to the previous case. The differences is that all the changes are much more violent. The occurrence of reverse waterhammer is much more likely than in other operations and Kaplan runners may be lifted by strong negative axial thrust. This must be checked by computation together with the calculation of the maximum values of x_i and Δh. If these results are unacceptable then alterations are necessary as stated above.

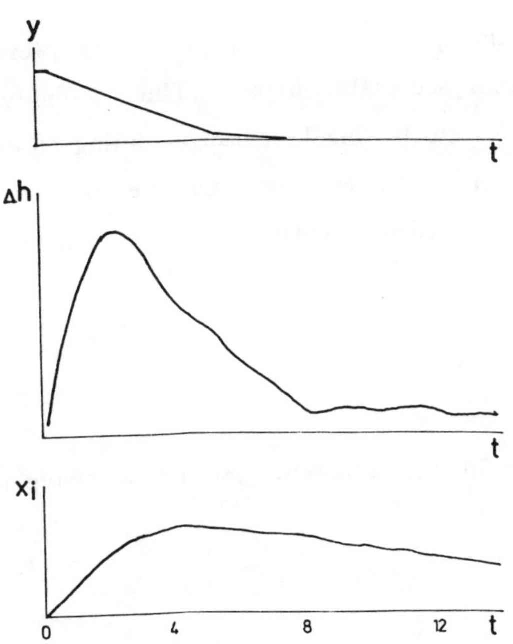

Fig. 6.12 Turbine quick closure

6.2.9 *Load rejection from pump operation*

This event is usually caused by a power failure disconnecting the unit from the electrical network. The wicket gates close quickly as shown in Fig. 6.13 causing the pressure in the penstock to drop and rise in the draft tube. At the same time the rotational speed of the runner is decreased. The process is frequently accompanied by strong oscillations, as shown in Fig. 6.14,(1) which should not be neglected in the design stage.

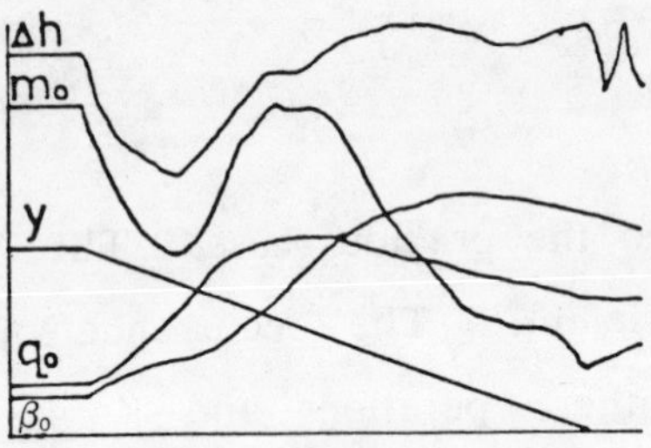

Fig. 6.13 Load rejection from pump operation

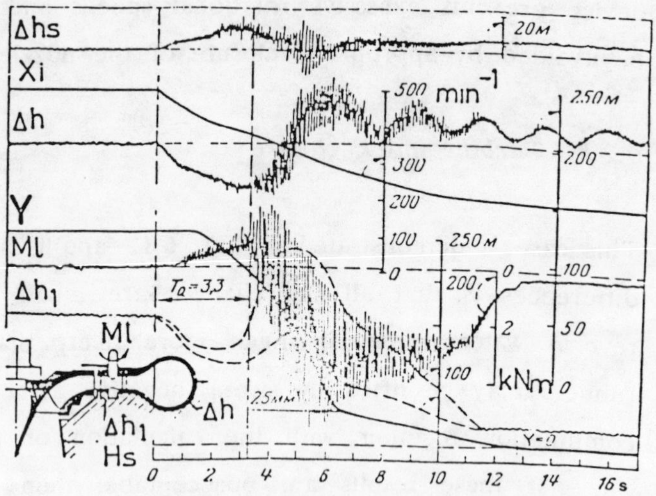

Fig. 6.14 Load rejection from pump operation; site recording showing strong oscillations

The computations will show maximum speed rise $x_{i\ max}$ and pressure rise Δh_{max} which should be kept within acceptable limits. The possibility of reverse waterhammer in plants with long penstock tunnel or runner lifting (Kaplan units) should be carefully checked. For machines with movable blades the movement of the runner blades during the transient regimes should be analysed as well.

6.2.10 *Hydraulic vibrations*

Many sources of forced oscillations exist in the apparently steady state regimes such as:

(i) vortices produced by the machine itself,

(ii) valves,

(iii) pipe bends,

(iv) enlargements.

Most of these oscillations are weak and could be neglected but there are important ones too. The problem is that there are too many possible sources acting in a wide frequency range which makes the task very complicated and incurs considerable expense. To obtain a reliable solution in a reasonable time at an acceptable cost, the analysis should be directed towards

(i) investigations if resonance may occur

(ii) computation of additional strains in the system due to pressure oscillations.

The first task is to compute natural frequencies (free vibrations) of the system over a wide range, by using either the matrix or impedance method. These should not fall near the frequency of any forced vibration, the most important ones being:

(i) vortex core in the draft tube,

(ii) disturbances due to rotation of the runner.

The vortex core is formed under the runner at the entrance to the draft tube. It rotates with a frequency given by[11,21]

$$f_s = \frac{1}{(2.8 \sim 7.6)} \quad \frac{n_o}{60} \quad \text{(Hz)} \tag{6.7}$$

depending on the operating point. The forces generated by this vortex may be rather strong, even if no resonance is present, especially at small loads. Oscillations of flow due to this phenomenon may equal 3 to 4% of rated discharge. The resonance should be prevented by air injection or other methods. The designer should be aware that air injection is not a universal answer since it may provoke resonance itself.

Mode shapes, especially maximum and minimum pressure in the whole system, should be calculated in the zone of frequencies of characteristics excitations. The analysis will show whether any protection device is necessary in order to diminish pressure fluctuations.

6.3 Emergency operating conditions

A hydro power plant is a very complex system with many devices. In spite of all the precautions, good maintenance and sound design, it is probable that one or other of the devices may fail sometime, either during a transient regime or in steady state operation thus creating a transient regime. Such an event constitutes an emergency case and the plant is generally submitted to higher strains than in normal transient situations but nevertheless should be able to withstand these and remain undamaged and operational.

The list of emergency cases is unlimited. The designer should select those which are more probable and more dangerous. A few characteristic cases are given below:

(i) machines with variable closing speed of the wicket gates. One servomotor mechanism may maintain the maximum closing rate to the end of the wicket gate closure, thus creating a shock in the system.

(ii) plants with synchronous relief valves. One of the valves may fail to open.

(iii) plants with several units. The wicket gates of one unit may remain open during the closing–down of the whole plant. This machine has to be closed by its penstock valve.

(iv) gate in the inlet building. In low head plants this gate may malfunction, either by closing down permanently or not completely.

(v) gate in draft tube. In some low head plants this gate is a protection against a runaway condition and may fail to operate.

(vi) runaway regime. This is caused by a machine remaining open after a power failure whilst the rest of the machines in the station close down normally. The water forces the open machine to the runaway zone regardless of the initial conditions. This dangerous state is terminated by closing down the inlet gates, or the draft tube gate, or the emergency spool valve closing the wicket gates.

(vii) unsuccessful start. Rapid start up of the whole plant followed by failure of one machine which has to be closed down immediately.

(viii) unfavourable set of events. Usually start up or closing down of the whole plant before the transients caused by an earlier operation had not died out resulting in amplification of the pressure surges. Initial and boundary conditions should be the most unfavourable. For example the water elevation in the surge tanks and/or air vessels should be either maximum or minimum depending upon which is worst for the given case.

The analysis should give extreme values of pressures throughout the system, maximum speed of runners, extreme values of water level in surge tanks and air vessels, etc. No reserve is needed to provide for in design of surge tanks and air vessels[22]. No water column separation is allowed in conduits (tunnels, penstocks).
The system should not be endangered by these excessive surges. Turbines, penstocks, tunnels, surge tanks and protection devices should be designed with a safety factor of 2 or 3 in regard to ultimate strength of the material[22] or 1.4 to 2 compared with the tensile strength [20].

6.3.1 Machine with variable wicket gate closure rate

Modern turbines usually have a two–speed wicket gate closure mode; at small openings the speed of the servomotor piston is decelerated in order to avoid high pressure surges. It is possible that due to the governor malfunction this deceleration does not take place, causing the wicket gates of one unit to close at maximum speed, as shown in Fig. 6.15. The resulting pressure surges are much stronger.

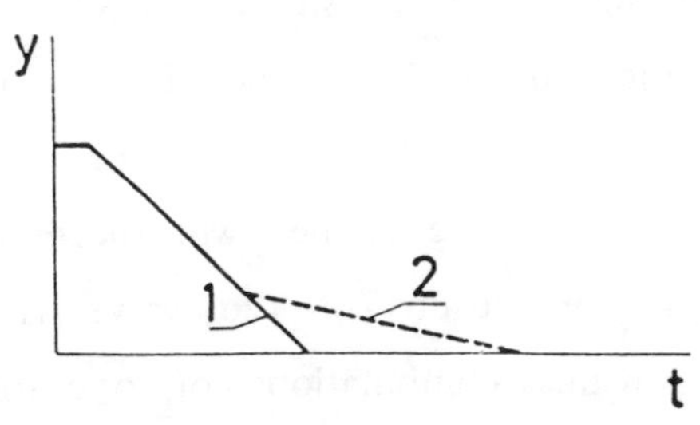

Fig. 6.15 Emergency operating condition. Two–speed wicket gate closure mode

The analysis should cover several cases with various combinations of operating machines and starting from different initial conditions. The worst combination giving maximum pressure surges is the criterion for design.

For pumped storage projects cases of load rejection from both pump and turbine operation should be analysed.

6.3.2 Plants with relief valves

The synchronous relief vaives divert part of the discharge from the turbines immediately after a load rejection thus relieving the system of high overpressures. The wicket gates may be closed at the maximum speed preventing high overspeeds. The arrangement is very efficient, but may not function properly when required. This is a common drawback of all devices which operate only intermittently. The analysis should cover cases when one valve fails to open whilst the remainder function properly, for various initial conditions.

The jet deflectors incorporated in Pelton machines are designed for the same purpose as the relief valves and therefore an analysis assuming their malfunction should be performed.

6.3.3 Plants with several units

The usual arrangement in plants with several units is that each unit has its own penstock valve situated upstream (turbining) direction of the unit. In some cases, during a rapid closure both the wicket gates and this valve are closed simultaneously as shown in Fig. 6.16. The resulting pressure surges depend to a certain extent on the timing of these two closures.(2) It is possible that the wicket gates of one machine fail to close (stuck in the fully open position) so the valve has to close the unit as shown by line 2 in Fig. 6.16. As a rule, pressure surges and/or maximum speed of the runner will therefore be higher than in the normal case, especially if the closure of the valve is delayed. Several cases should be analysed for various combinations of operational units and different initial conditions assuming one unit fails to close the wicket gates.

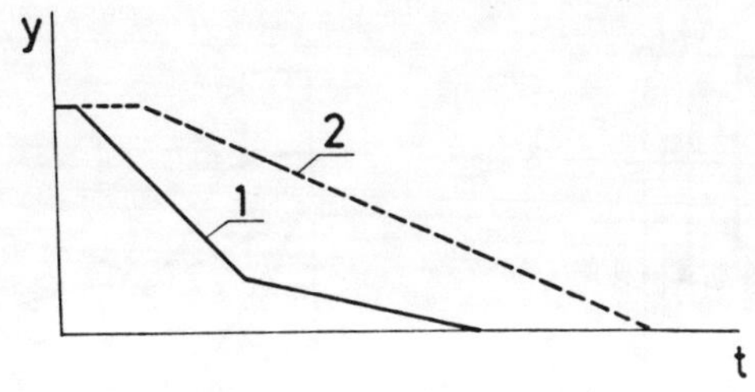

Fig. 6.16 Simultaneous closure of wicket gates and penstock valve during emergency rapid closure

6.3.4 Gate in the inlet building

In low head plants the inlet gate may create problems during the closing–down and emptying of the plant. It can happen accidently that the wicket gates close down when the penstock is already partly emptied and the gate almost closed. Large quantities of air are then trapped and serious pressure surges will follow. Dangerous conditions may also appear during the filling, especially if the gate is lifted more than anticipated.

The analysis should determine proper timing and sequence for the various actions during the closing–down of the whole plant. Simultaneous closure of the inlet gate and wicket gates is not recommended.

6.3.5 Gate in the draft tube

A gate located in the draft tube sometimes replaces the one in the inlet building resulting in a cheaper layout.[11] This is shown in Fig. 6.17 together with the time–dependent variations in the various parameters during the closure mode. The gates are controlled by oil servomotors, activated by different means. The speed of lowering the gate is determined in such a way that the maximum speed of the rotor is not too high whilst keeping pressure surges as low as possible. Special attention should be given in order to prevent the negative axial force which may lift the runner.

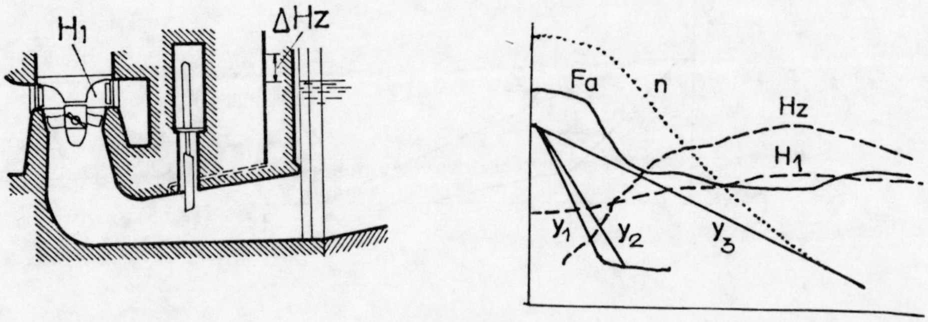

Fig. 6.17 Emergency closure of the gate in the draft tube

This system is, up to now, only applied to plants with Kaplan units. The analysis should therefore be made for one unit only, but for several initial steady state regimes.

6.3.6 Protection against runaway conditions

Special precautions are usually taken to ensure the closing-dowr of the wicket gates. For instance, if one or two blades are stuck their levers will break allowing all the other blades to close normally. However, the possibility of all the wicket gates malfunctioning cannot be totally ruled out and therefore there are always additional devices which will prevent runaway of the unit if the wicket gates fail to close. These are

(i) A valve immediately upstream of the machine (Pelton, Francis, Kaplan or diagonal flow types),

(ii) a gate in the inlet building (Francis, Kaplan or bulb type),

(iii) a gate in the draft tube (Kaplan),

(iv) an emergency spool valve which connects directly the oil tank with the servomotor moving the wicket gate apparatus.

These devices constitute additional safety and are absolutely necessary in order to prevent water flowing through a damaged unit. When the protection devices are operating normally the unit should be able to restart without any additional control.

The devices mentioned above are activated by a sensor detecting overspeeds 10 to 15% higher than the rated overspeed caused by the servomotor speed being slower than the computed speed.

The second stage of protection is a device which is activated when the rotational speed exceeds a given value of x_{ir}. If the maximum runaway speed after a power failure is given by

$$x_{imax} < 0.6$$

then the device is set on

$$x_{ir} = x_{imax} + 0.05$$

Alternatively if

$$x_{imax} > 0.6$$

then the device is set on

$$x_{irmax} > 0.4 \text{ to } 0.5$$

with obligatory control of the speed of the runner. If the rotational speed is not diminished with time the protection is activated.

The rotor of the electrical machine should be able to withstand a speed 1.75 times the rated speed for two minutes(9). Temporary overspeed in such conditions should be less than $x_{imax} = 0.6$, (see Chapter 9.2).

6.4 Catastrophic cases

There are disastrous cases when the plant is endangered by simultaneous failure of two or more safety devices, or by an extremely unfavourable series of events, or by natural forces such as an earthquake or flood. Some of these cases are described below.

(i) Machines with variable closing speed of the wicket gates. Two or more servomotor mechanisms maintain the maximum wicket gate closing speed over the closure period.

(ii) Plants with relief valves. Two or more relief valves fail to open during a rapid closure.

(iii) Plants with several units. The wicket gates of two or more units remain stuck in the fully open position; the corresponding penstock valves close normally.

(iv) Runaway. One or more units go to runaway because all devices in their branches fail.

(v) Instantaneous closure of wicket gates. Wicket gates close in a time less than 2L/a.

(vi) Instantaneous closure of a valve or gate.

(vii) Column separation and reverse waterhammer.

(viii) Resonance in the wicket gates. Oscillatory movements of its blades with the frequency equal to one of the natural frequencies of the system.

(ix) Auto–oscillations provoked by leakage on an imperfectly closed valve. Usually caused by a ball valve.

(x) Breakdown of the penstock. For instance a crack in the penstock near to the powerhouse.

(xi) Earthquake. Water within the system will act as a piston during an earthquake.

The pressure surges produced by such unlikely events are, as a rule, too strong to be taken as the criteria for design, especially if their probability is extremely low. However the analysis should cover these cases too in order to ascertain what will happen to the plant in the case of such an event. The strains within the system should not exceed the yield point or the ultimate bursting strength of the material, if this does not lead to unacceptably high costs. If the damage cannot be prevented, then its extent should be limited by installing rupture discs and similar self–sacrificing devices, for instance.

The damage caused by a catastrophic event is treated as a 'vis major' situation if there has been no human error, (such as inadequate maintenance or disobeying explicit rules).

6.4.1 Machines with variable closure speed of the wicket gates

In the case of a two–or three–speed closure mode of the wicket gates, the maximum speed of the wicket gates may prevail for two or more machines simultaneously until the gates are slammed shut.

6.4.2 *Plants with relief valves*

Cases when two or more valves do not function properly should be analysed in order to determine pressure envelopes along all the conduits.

6.4.3 *Plants with several units*

The wicket gates on two or more units fail to close, therefore these branches are closed by their valves or otherwise (gates in the inlet or draft tube). The analysis should cover all the cases possible in reality. The main results are pressure envelopes along all the conduits and the maximum temporary speed of the runners.

6.4.4 *Runaway*

The full runaway regime cannot be prevented for all conditions, therefore means of stopping the machine must be provided. The analysis should determine the maximum runaway speed and the associated pressure surges.

6.4.5 *Instantaneous closure of the wicket gates*

This may occur as a result of damage within the control system. If the wicket gates close in a time less than $2L/a$ then a serious shock will appear with an amplitude given by

$$\Delta h = \frac{ac_o}{g} \tag{6.8}$$

where a = wave speed (ms^{-1})

c_o = initial water velocity (ms^{-1})

g = acceleration due to gravity (ms^{-2})

In the opposite case, when considerable column separation occurs the pressure surge amplitude may be even higher. The analysis should determine the pressure envelopes along all the conduits. The possibility of reverse waterhammer occurrence, or generally water column separation anywhere, should be carefully analysed.

6.4.6 Instantaneous closure of a valve or gate

This may result due to a failure, human error, or breakdown of equipment. The analysis should deal with reasonably possible cases, consulting experts whenever possible. The envelopes of pressure should be determined for all conduits.

6.4.7 Column separation and reverse waterhammer

The occurrence of a vacuum is never advantageous because it may cause strains which the system has not been designed to withstand. If the local pressure falls below the vapour pressure then dissolved gases are released, water evaporates and a cavity is formed. From that point the numerical analysis is based upon several hypotheses none of which has been definitely proved to date. This event is very dangerous because it may cause excessively high pressures and even destroy the penstock or damage the machines. In plants equipped with Kaplan or bulb type machines the reverse waterhammer may appear in the draft tube. Firstly a large cavity is formed under the runner and later when the cavity collapses the water from the lower pool rushes in, hitting the runner hard. Cases have been reported where this shock was strong enough to lift all the rotating parts causing severe damage. This is likely to happen in any plant with a long outlet tunnel.

The analysis should determine if the reverse waterhammer may occur or not. There is no point in trying to design a system strong enough to withstand these shocks; the design should be oriented towards preventing such an event.

6.4.8 Resonance caused by wicket gates

The governor may start to 'hunt' variations of load, i.e. to open and close the wicket gates periodically as shown in Fig. 6.18. This may happen if the governor is out of order or badly tuned. The analysis has to determine frequencies of free vibration of the system by using a linearised mathematical model. Then oscillations of pressure due to vibrations with these frequencies should be computed, taking into account the characteristics of the governor. If the resonance is likely, then full waterhammer equations should be used. Such a case should be prevented.

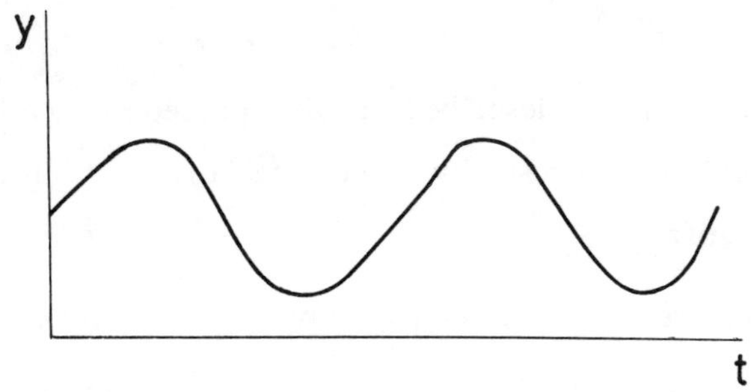

Fig. 6.18 Periodic opening and closing of the wicket gates due to the governor hunting variations of load

6.4.9 Auto–oscillations in the plant

Auto–oscillations of the water masses in a hydro power plant appear relatively often and when they do appear the plant could be severely damaged. The most common cause is the leakage through an imperfectly closed valve, usually a seal ring in a ball valve not being kept tight enough. Periodic flow through such a ring will bring about resonance, given enough time. The analysis should compute the first three harmonies of the system and find out its response. The linearised theory is normally used at the beginning but if the case is a difficult one, the full waterhammer equations can be applied.

6.4.10 Earthquake

In the event of an earthquake all tunnels, penstocks and outlets will be suddenly moved, and serious waterhammer effects created. Pressure surges in conduits should be computed for characteristic cases (typical earthquakes). The most dangerous situation is usually when all machines and valves are closed and the principal movement of the ground is parallel to the axis of the conduit with a dead end.

The strain should not exceed the ultimate strength of material even in the worst case. However, if this leads to excessively high costs then it may be ignored since the probability of the very worst case is very low.

6.5 Presentation of transient regimes

The results of all the analyses described in the preceding sections are voluminous tables of digits difficult to digest by non–specialists. The results should be presented in the form of:

(i) time variation for selected machines, valves, surge tanks etc.,

(ii) pressure envelopes for all conduits,

(iii) the path of the transient on the machine performance characteristics.

6.5.1 Turbine operation

A turbine may be either operating in the zone of turbining (A) or the zone of braking (B) as shown in Fig. 6.19. These two zones are separated by the line of zero torque corresponding to zero efficiency.

Several transient regimes are shown in Fig. 6.19 for a Francis type turbine.

(i) line 1 : full runaway of the turbine after a power failure and subsequent stoppage.

(ii) line 2 : application of emergency valve to cut short runaway development followed by stopping the machine.

(iii) line 3 : rapid closure of a turbine after a power failure.

(iv) line 4 : load acceptance.

(v) line 5 : normal stoppage.

(vi) line 6 : normal start–up.

The same transient regimes are shown on the $\varphi - \psi$ diagram in Fig. 6.20.

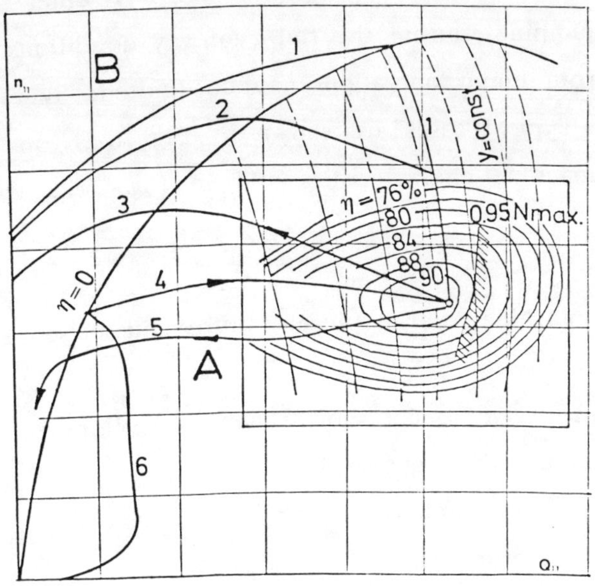

Fig. 6.19 Transient regimes superimposed on the unit parameter performance characteristics of a Francis type turbine

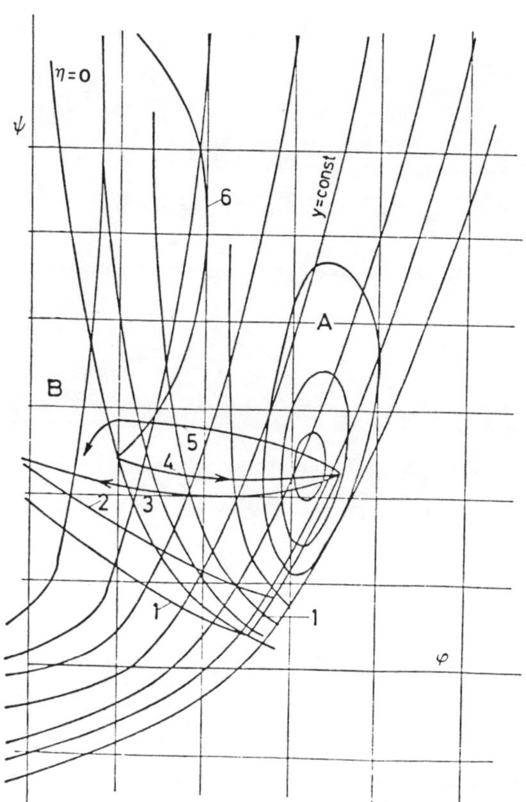

Fig. 6.20 Transient regimes superimposed on the dimensionless parameter performance characteristics of a Francis type turbine

6.5.2 Pump operation

Transient regimes of pump turbines in turbine operation are similar to those described above with one important difference; a pump turbine may enter into the fourth quadrant (called reverse pumping) during the full runaway conditions.

Transient regimes which start from pump operation may be limited, by rapid closure, to the first two quadrants. Typical cases are shown in Fig. 6.21 on the unit parameter presentation of the machine characteristics and in Fig. 6.22 using $\varphi - \psi$ coordinates.

(i) line 1 : full runaway of a machine after a power failure and subsequent stoppage.

(ii) line 2 : closure of emergency valve to cut short runaway development after the machine has entered the third quadrant.

(iii) line 3 : rapid closure after a power failure.

(iv) line 4 : start−up in pump operation.

(v) line 5 : stoppage from pump operation.

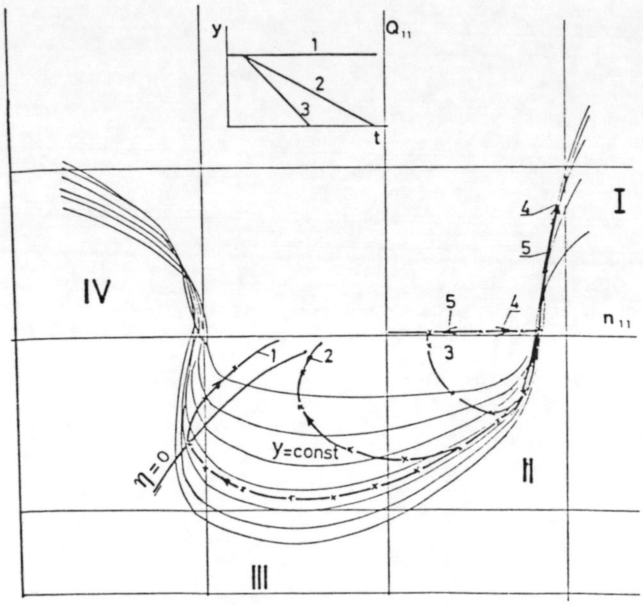

Fig. 6.21 Transient regimes superimposed on the unit parameter performance characteristics of a reversible pump turbine

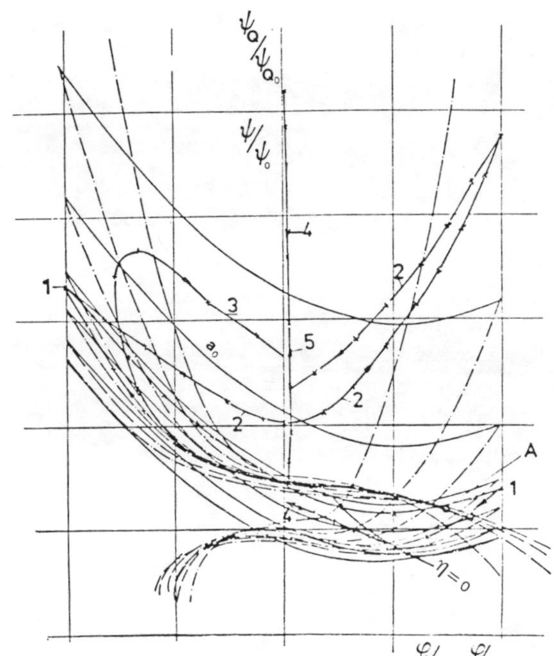

Fig. 6.22 Transient regimes superimposed on the dimensionless parameter characteristics of a reversible pump turbine

7 Analysis of transients

7.1 Progress of analysis

For any one hydroelectric installation the transient regimes are analysed, as a rule, at the following stages:

(i) during the feasibility study,

(ii) during the general design,

(iii) during the detailed design (after bidding),

(iv) during the commissioning and running–in process,

(v) during any reconstruction or enlargement designs,

(vi) during trouble–shooting investigations.

It is obvious that the degree of accuracy and reliability of the data, together with the complexity and accuracy of results, increase as one progresses down the above list. For instance, in the feasibility study, it is important to determine whether any waterhammer protection is necessary for the installation or not, and to select and compute all important cases referred to in Chapter 6. For such analyses it is possible to use appropriate data available from the literature. On the other hand, if something goes wrong with an existing installation special experiments may be necessary in order to determine the exact value of various parameters, such as the wave velocity in different parts of the system, so that the analysis can accurately demonstrate the responsibility for the damage.

Typically the first analysis of transient regimes is performed as an integral part of the feasibility study. The data are compiled on the basis of previous experience or taken from the literature. The analysis is centred on normal and emergency cases, with the aim of determining the best design for the future installation. Different installation layouts are normally analysed.

More data are available at the general design stage and the analysis of transients is extended to incorporate some of the catastrophic cases. The results of this analysis assist the engineers in improving the design of the various elements of the plant. This underlines the necessity for good teamwork between all the designers in order to obtain a successful outcome.

The results of transient analyses are usually reported in a separate volume of the general design, and used for the preparation of the tender documents.

Up to this stage the transient analyses have been an integral part of the design procedure, helping the designer to formulate the best solutions and to avoid hidden dangers for the future plant. Now the manufacturer becomes involved. Usually he makes a few essential computations of transient regimes, using experience and data relating to his machine and equipment, when preparing his tender bid. Having obtained the contract, the manufacturer proceeds with more detailed computations, trying to reach solutions which would best suit his equipment. At this time, the designer is engaged in the detailed design. Obviously good cooperation between the two is very important but some degree of friction is inevitable. The client should not make the mistake of leaving the analysis of transient regimes to the manufacturer who may offer to do so 'free of charge'. Even the most honest manufacturer is prone to use results to prove that his machines and equipment are excellent and delete hints of possible future troubles. Therefore the analysis of transient regimes using the manufacturer's data must be repeated. These results are usually presented as a separate study or as a part of the general design. All normal, emergency and catastrophic cases are to be analysed because this is the last chance of correcting serious mistakes.

The main findings of all this work are checked during the commissioning and running–in of the plant. Special measurements of transient regimes are recorded and the results compared with predictions. However, this applies only to *normal* cases, very rarely to a few *emergency* cases and certainly to none of the *catastrophic* cases. No client would, or should, take chances with a powerful and very expensive project. This means that only the most harmless events will be checked and not the most dangerous ones. Nevertheless these investigations are of paramount importance for all concerned with the installation. The agreement between computed and measured values should be good. Any discrepancy should be carefully traced to the source and explained. In that case the calibration of the mathematical model according to the measured data should be performed and all computations must be repeated.

The reconstruction or enlargement of an existing plant also begins with the feasibility study and follows the same steps. Very often transient regimes are more dangerous in the reconstructed or enlarged plant than in the existing one. So a new, very detailed study of transients must be undertaken as opposed to just extrapolation of the findings of the available design documentation. This new study should cover all normal, emergency and catastrophic cases.

The problems associated with transient states may arise in operational installations, sometimes even after several years of undisturbed operation. These problems usually relate to emergency or catastrophic cases, damage to the plant due to breakdown of equipment, excessive vibrations or other similar situations. To determine the exact cause of the trouble and to find an adequate cure, a very detailed and precise study should be undertaken.

Small hydro power plants, especially those incorporated in larger systems, suffer fom the same problems as the large ones. The analysis of transient regimes may be much more complicated, due to complex boundary conditions. On the other hand, the general tendency is to decrease the cost of design and, therefore, to simplify the analysis.

7.2 Selection of parameters

The analysis of transient regimes is based on data which are positively known and others which are assumed. Initial and boundary conditions also vary over a wide range, so even the most extensive analysis can only cover a few, presumably characteristic, combinations. These assumptions regarding data, initial and boundary conditions constitute parameters of the analysis. At the beginning, during the feasibility study, there are a lot of assumptions. Later, as the work progresses, assumptions are replaced by facts and the number of parameters decreases.

All parameters should be listed at the beginning of a report on the analysis of transient regimes. A sensitivity analysis should be made on the most important parameters, some of which are discussed below.

7.2.1 *General layout of the plant*

Any analysis starts from the layout of a plant. A plan view of the entire plant, a longitudinal cross–section and plans of the machine house, surge tanks, inlet and outlet buildings should be prepared as precisely as possible at the given stage of the project design. The exact locations of equipment such as valves, gates and air valves, should be noted on the plans.

7.2.2 *Tunnels, penstocks and pipelines*

The length, diameter and position of tunnels, penstocks and pipelines have to be determined before the analysis of transient regimes may commence. Construction details, thickness of pipe walls, characteristics of the materials and pipe anchoring are also essential information for such an analysis. However, in the feasibility study the values of these items of data may not be fixed and are therefore parameters of the analysis.

If at any stage there is a significant change of any of the above parameters the computation should be repeated. Analysis of hydraulic vibrations is particularly sensitive to these parameters.

7.2.3 *Water levels in the storage basins*

The water level in the upper storage basin usually varies over a broad range, depending on factors which are not within the scope of this book. For the analysis of transient regimes it is essential that complete information regarding the possible range of water level in the storage basin is provided, including abnormal cases such as operation during the construction period or freezing period.

The level in the lower pool may depend on various factors such as the discharge through the plant, operations of a downstream plant, or the dredging of the tailrace water basin.

The minimum water elevation in the lower basin is assumed for all the analyses of possible reverse waterhammer since this corresponds to the lowest average pressure below the runner.

7.2.4 *Analysis of stability*

The stability of the regulation process is not covered in this book but such an analysis is essential and should be performed separately. It is partly based on the same data as computations of transient regimes and should embrace the whole system.

7.2.5 *Pressure losses*

There are two types of losses, friction and local. The frictional losses depend on the internal roughness of the pipe or tunnel walls and vary with the age of the installation. The ageing cannot be exactly estimated in advance because it is a complicated process depending upon the chemical, physical and biological properties of the water. The most important cases should therefore be analysed for the minimum and maximum values of the friction coefficients.

7.2.6 *Initial and boundary conditions*

The results of computation are a function of the initial and boundary conditions. There is an unlimited number of possible combinations, even if all the main parameters are fixed, but the time and resources are always limited. The analysis should include only the characteristic, relevant, cases which will provide all the necessary information for further design processes. Selecting the appropriate cases is not straightforward since the following criteria have to be considered:

(i) cases which produce extreme values of pressure surges, machine speed and variations in water elevations.

(ii) only the cases which are possible in reality.

It is obvious that these two criteria sometimes disagree: a highly improbable case might be the most dangerous one. The designer has to make his choice and justify it. The recommendations listed below are provided in order to assist in these decisions:

(i) analyse cases with extreme values of discharge and head.

(ii) analyse cases when only one unit is operating (start–up, stopping, change of load, power failure) and also cases when all the units are operating (simultaneous power failure, emergency cases).

(iii) analyse cases resulting from failure of an important device.

(iv) analyse unfavourable sets of events (i.e. power failure during a start–up at the most inopportune moment).

7.2.7 *Characteristics of the governor*

The way of opening–up and closing–down the units, expressed through the function y(t), has a paramount influence on transient regimes. A typical closure relationship was shown in Fig. 4.1 indicating the dead time (T_q), the minimum closure time (T_f), the position (y_h) from which the movement of the piston is decelerated and the interval of the decelerated closure (T_h). Two regulation systems are frequently employed.

(i) the shape of the starting/closing process is always the same regardless of the initial and boundary conditions.

(ii) the maximum speed of the servomotor piston and mode shape are adjusted according to the program so as to decrease maximum runaway and maximum and minimum pressures in the conduits.

The control variable in the second case above is a measured parameter; either a local pressure, the head, or the appearance of a vacuum. A prearranged program changes the speed of the servomotor, depending on the actual value of the control variable, trying to strike the optimum balance between transient overpressures and overspeed which will follow. The life of a plant and its vital parts will be much prolonged this way. Any damage of this device may provoke serious pressure surges. This event is classified under emergency cases.

The analysis of transients should be made for maximum servomotor speeds, which corresponds to extreme positions of the spool valve[9]. To take into account effects of changing quality of oil, ageing of oil, seals, temperature and other influences, the analysis should be carried out with:

(i) shortened time, $T_{f\ comp} = 0.9\ T_f$, when maximum pressure surges are sought.

(ii) prolonged time, $T_{f\ comp} = 1.1\ T_f$, when maximum temporary overspeed is being looked for.

The dead time (T_q) may also be varied, but within reasonable limits.

The values of y_h and T_h should be determined so as to ensure that neither pressure waves at the end of closure are stronger than those in the first phase of the process, nor any of the later shocks, when reflected waves hit the closed machine[9].

The computations performed in the early stages of the design (feasibility study, preliminary design) should be made with linearised functions taking the dead time (T_q) into account $y(t) = 1 - t/T_f$ or $a(t) = a_o(1 - t/T_f)$ where $a(t)$ is the wicket gate opening taking the dead time (T_q) into account.

7.2.8 *Characteristic diagrams of the units*

Four–quadrant performance characteristic diagrams (see Chapter 5) should be based on measurements on a small–scale model. They must be used for final computations. Scaling effects are usually disregarded in the analysis of transient regimes, as well as possible effects of non–stationary flow within the machine upon its performance.

If such data are unavailable and the designer has only a portion of the hill diagram around the point of best efficiency, then four–quadrant diagrams have to be estimated using these data, experiences and curves for similar units. The accuracy of computation may be decreased considerably due to this extrapolation.

7.2.9 *Injection of air in the draft tube*

Vibrations caused by a vortex core, formed below the runner, are sometimes attenuated by injecting air into the draft tube. This may lead to a higher runaway speed at partial loads. The model performance curves are not reliable in this case because the similarity laws are not satisfied due to the two-phase flow.

Air injection also changes the frequencies of free vibrations as well as the amplitude and frequency of vibrations caused by the vortex core. This should not be neglected in the anlysis of transient regimes or stability of regulation.

7.2.10 *Inertia of rotating parts*

The inertia of all the rotating masses (runner, shaft, coupling, rotor of electric machine, water within runner blades) has a considerable influence on transient states. This effect is especially important when analysing the consequences of a power failure during pump operation.

In the early stages of the analysis (feasibilty study, preliminary design) a sensitivity analysis should be performed as soon as possible to determine the appropriate value of the inertia.

The ratio of the starting time of the water masses (T_W) and the mechanical starting time (T_a) is another parameter which has a great influence on transient states and should be investigated in the early stages of the analysis.

7.2.11 *Relief valve*

The characteristics of such a valve (flow coefficient against valve opening) are usually determined on a small-scale model. For the first computation, this data are not available so information relating to similar devices should be used instead. Timing and maximum opening or closing rates are among the parameters which should be determined during the analysis.

7.3 Results

The analysis of transient regimes should achieve the following:

(i) calculation of the extreme values of pressure, machine speed and water levels together with associated fluctuations.

(ii) determination of the best parameters for regulation; opening-up and closing–down of the machines and valves, etc.

(iii) selection and design of adequate protection devices and installations.

(iv) evaluation of potential dangers for the plant during any unsteady conditions.

(v) establishment of operational rules for the plant in emergency and catastrophic events.

The results of the transient analysis should be presented in a separate report, according to the general rules for the technical documentation for the given installation. The minimum requirements for such a report are listed below.

7.3.1 *Conduits (tunnels, penstocks, pipes)*

The results should include

(i) envelopes of maximum and minimum pressures along the conduits, at least for the most characteristic cases.

(ii) amplitudes of pressure pulsations due to forced vibrations of the system (the most probable source).

(iii) first three harmonics of the self exciting oscillations of the system with amplitudes of free vibrations drawn for all conduits.

Special attention should be given to eventual dangers such as (a) the appearance of a vacuum, especially if cavities are formed and water column separates, followed by reverse waterhammer and (b) the appearance of resonance in the system. Both of these cases should be avoided if possible, except in the catastrophic situations.

7.3.2 *Surge tanks or air vessels*

The results should include:

(i) maximum and minimum pressure in the tunnel and water levels during the transient states.

(ii) time history for the most characteristic cases.

Special attention should be given to events such as water spilling over the weir, vacuum occurring in the tunnel, the surge tank emptying or an air vessel subjected to overpressure or vacuum. Generally such cases should be avoided because the surge tank or air vessel may be seriously damaged.

7.3.3 *Turbines with fixed blades*

The results should include:

(i) parameters relating to rapid closure which are finally accepted (i.e. T_q, T_f, T_h, y_h; Fig. 4.1).

(ii) maximum temporary pressure rise and speed of rotation of the runner for a few characteristic cases.

(iii) parameters of opening−up and closing−down procedures (with speed equal to the rated value).

(iv) universal characteristic of regulation as shown in Fig. 6.7[9,11].

(v) time history for a few characteristic cases.

(vi) runaway speed of the machine under the most critical conditions.

If the installation has more than one unit connected to the same penstock then mutual influence of machines should be investigated and results presented for cases such as:

(i) temporary changes in the operating machine as a result of start−up, closure, change of load, or load rejection of other machines.

(ii) runaway state of one machine, the wicket gates of which remain fixed in the fully open position, after a simultaneous load rejection of all operating units, all other units being closed normally.

Other cases are mentioned in Chapters 6 and 7 as well as Appendix III.

Special attention should be given to cases when extreme values of pressure surges and runaway speed are not within prescribed limits, when a cavity is formed under the runner with a consequent reverse waterhammer, or when a vacuum appears in the spiral casing. Such cases are, naturally, to be prevented if possible.

7.3.4 *Turbines with movable blades*

The results should be the same as those listed above for turbines with fixed blades with the additional parameter of the angle of the runner blades. Special attention should be given to the analysis of reverse waterhammer which is more probable in low head plants. The computations are usually made for one machine only, because common conduits are rare in such plants and therefore there is no mutual influence between machines.

7.3.5 *Pump turbines with fixed blades*

The results should be the same as those listed above for turbines with fixed blades plus additional results for pump operation modes.

It is important to analyse the impact of the pipeline inertia (T_W) and machine inertia (T_a) on transient regimes in pump operating modes at an early stage in the design when penstock length, diameter and position, inertia of rotating parts and other parameters are to be determined.

Special attention should be given to investigation of transients if the machine tends to enter the fourth quadrant (reverse pump operation). The case when power failure occurs to all the machines operating in turbine mode and the wicket gates of all the machines are jammed in the open position should be analysed as the most dangerous case in that respect. This is a catastrophic case, highly improbable in any condition, but which reveals if the machine may enter the fourth quadrant. This phenomenon is accompanied with strong pressure surges caused by the 'S' shape of the characteristic curves in the zone of runaway.

7.3.6 *Pump turbines with movable blades*

The results should be the same as those listed above for turbines with movable blades and pump turbines with fixed blades.

7.3.7 *Penstock valves*

The results for the valves located upstream of the turbines should include:

(i) parameters for rapid closure of the valve.

(ii) extreme pressures upstream and downstream of the valve.

(iii) time history for characteristic cases.

(iv) effects on turbines or pump turbines in cases when their wicket gates fail to close on time.

7.3.8 Relief valves

If such a valve is to be used for regulating flow during transient regimes the results should include:

(i) parameters for opening-up and subsequent closing-down (Fig. 6.11).

(ii) extreme pressures in the system and the maximum flow through the valve.

(iii) time history for characteristic cases.

(iv) effects on the turbine.

7.3.9 Gate in the draft tube

If this gate is to be used for regulation of flow during the transient regimes, the results should include:

(i) parameters for rapid closure.

(ii) extreme values of pressures in the system.

(iii) time history for characteristic cases.

The effect on the turbine should be carefully investigated.

7.3.10 Special considerations

There are dangers for the plant which should be separately investigated, because they may appear in an event not included in other computations. These are:

(i) protection against runaway. A safety device which may fail or act at an inopportune moment.

(ii) reverse waterhammer when air is admitted through the head cover. Partial load cases are sometimes more dangerous than those starting from full load.

(iii) lifting of the runner by strong negative axial force; again partial load cases may be more dangerous.

If there is any hint that such a disaster might happen, then a complete and well documented chapter of the report should be devoted to this subject. If there is any suspicion that any transient regime can cause damage to the plant a special analysis should be performed and the report should include all possible details.

account, a [illegible] of the [illegible] should [illegible] and [illegible] effectively [illegible] connection [illegible] centred [illegible] to cause damage to [illegible] analysis should [illegible] and the [illegible], should [illegible]

8 Measurement and investigations

8.1 Classification

Investigations are performed at the following stages for any given installation:

(i) during trials and running–in period,

(ii) during the commissioning procedure,

(iii) whenever operational troubles and minor damages tend to accumulate,

(iv) after an accident resulting in damage to the plant,

(v) at regular intervals.

The aim of investigations carried out during trials is to find out if the machines and associated plant can operate safely, avoiding any possible danger to the plant and personnel. Investigations performed during commissioning have to prove that the plant can attain the desired performance in normal operation. Investigations of unusual behaviour of the plant have to show what is the cause of troubles in operation and, if possible, to help eliminate the cause. The main purpose is to prevent a possible accident. Investigations undertaken after an accident have to prove that the damage is repaired and that the plant may continue to operate safely. Regular investigations done as a part of the normal operational procedure should indicate whether the plant is functioning properly and thus provide the operators with an early warning before an incident occurs.

The measurements and investigations which are referred to in this chapter are related to the transient regimes. More details on general measurements may be found elsewhere[21].

The results of investigations into transient regimes should be carefully compared with the results of computations made during the design procedure. The agreement between measured and computed values should be good. Any discrepancy must be analysed and explained, whether it is caused by the inaccurate data or the physical system is different from the mathematical model. The calculations should be repeated until a satisfactory agreement is achieved.

If an accident has occurred in spite of all precautions, such a case should be examined in detail, in order to discover the cause and to use this bitter experience to avoid similar occurrences in future. A special program of measurements should be prepared for recommissioning of such plant.

Since the strains within the plant are at their maximum intensity during transient regimes, these investigations show the degree of safety of the plant.

8.2 Scope of measurements and investigations

As a rule, all normal regimes are investigated but only a few of the emergency cases and none of catastrophic situations.

8.2.1 Normal operating regimes

All normal operating situations should be tested during trials, but in such a way that the safety of the plant is never in danger. Therefore the tests start with the transient regimes which cause the smallest hydrodynamic loads, and then move towards more dangerous ones. The results of computations should be made available so that measured data could be instantly compared with the computed values. Any serious disagreements between the observed and computed data is a warning which should be studied carefully before the further investigations were resumed.

For instance, the investigations of load rejection cases starts with initial load of 25% only, then 50, 75 and finally 100% of rated power.

In the case when observed and predicted values fit well together, the program of measurements and investigations could be revised and shortened.

8.2.2 Emergency cases

The number of emergency cases which are investigated during trials is usually very limited in order to avoid any damage to the plant. Therefore only such cases which cannot provoke the maximum strains of the system are simulated in the plant, selecting accordingly the initial and boundary conditions. An example is the closure of the penstock valve after the wicket gates have failed to close down the machine (following a load rejection), but starting from 25% or 50% load.

If the agreement between the observed and computed values is good, further investigations are cancelled in most cases. In the opposite case, the reasons for disagreement should be carefully analysed. One reason might be unreliable data and assumptions on which the mathematical model was based. In that case the model should be calibrated and all computations repeated, until a satisfactory agreement is obtained.

8.2.3 Catastrophic cases

These situations are normally avoided and all decisions and precautions are based upon the theoretical investigations. Good agreement between computed and measured data obtained for normal and emergency cases is a guarantee that the computations are valid.

In very special circumstances such investigations might be conducted with the special permission of the higher authorities. Special precautions should be provided to prevent the damage (or a disaster), but if it happens nevertheless it is considered as a 'vis major' case.

8.3 Scope of measurements

Many important parameters are measured, continuously or intermittently. Here are listed only those parameters which are important for the transient flow analysis.

8.3.1 Continuous measurements

Continuous records on a suitable medium (paper tape, magnetic tape, diskette, etc.) are kept for:

(i) pressures in the chosen places (sites) along the tunnel, the penstock, immediately upstream and downstream of the machine, along the outlet tunnel,

(ii) pressures within the machines: in spiral casings, head cover, under the runner, in the draft tubes,

(iii) wicket gate openings,

(iv) angles of the runner blades for the machines with movable blades,

(v) strokes of the valves used for the control transient regimes penstock, such as valves,

(vi) speed of machines,

(vii) torque acting on the coupling,

(viii) axial hydraulic thrust.

(ix) displacement and vibration of bearings.

The data recorded in a proper way are very useful for all later analysis and investigations. The strip chart on the paper tape provides the information for analysing the measuring results at once, so decisions can be made whether the further investigtions could be dangerous or not. The purpose of these analyses is to discover a potential threat to the safety of the plant and to prevent it in time.

8.3.2 Intermittent measurements

For steady−state regimes before and after the transients, the following parameters are recorded:

(i) water level in the storage basin and in the lower pool,

(ii) head,

(iii) wicket gate openings,

(iv) angle of runner blades (for machines with movable blades),

(v) pressures in the penstock, upstream and downstream of the power house, in the outlet tunnel,

(vi) pressures within the machines: in spiral casings, under the runner, in the draft tube,

(vii) electric current and voltage in the generator (or motor—generator),

(viii) angular speed of the machines.

The data are recorded in the log book, and should agree with the corresponding values recorded continuously. Other information might also be recorded in these books.

8.4 Processing of the measured data

The data should be shown in the shape of a time—history diagram in a suitable scale, so that the comparison with the results of computation is easy.

It is extremely important to describe clearly the initial steady state regime as well as the situations which had developed afterwards (i.e., a load rejection, start, etc.). The time scale and also scaling factors for all physical parameters shown in the diagram should be clearly stated. The final steady state regime should also be given.

For analysis of hydraulic oscillations, diagrams of frequency domain should also be prepared in order to help the easy detection of the resonant frequencies.

8.5 Recommended tolerances

The differences between the observed and the computed values should not be greater than 5%, or for less accurate analysis greater than 10% of the maximum values. These tolerances are based upon the current 'state of the art' in this field and are quite arbitrary. Generally accepted rules do not exist, yet, and therefore it is advisable that the final analyses of transient regimes are done by the specialists, especially in the more important cases.

9 Recommended range of parameters

9.1 General

Maximum ranges of certain parameters, characteristic of transient regimes, are listed below. These values correspond to usual practice and the designer should try to use their full extent, staying within these limits whenever possible. However there are also numerous exceptions due to specific technical, economical or operating conditions of particular installations.

9.2 Speed of the runner

The rotor of an electrical machine can withstand speeds up to 1.75 times the rated speed for two minutes without lasting deformations. Therefore maximum temporary overspeed resulting from a power failure given by $x_{i\ max} = 0.6$ is permitted for standard hydro power plants. After a power failure the machine may restart without any examination providing that the above limit was not exceeded.

In special cases a greater value of overspeed is permitted, temporarily, but only with the consent of the manufacturers of the hydraulic and electrical machines and equipment. Such a request must be defined in the design and contract specifications.(9)

9.3 Protection devices

Both the governor and the special protection devices preventing the machines going to runaway should be tuned to act when the speed exceeds the permitted value as recommended in Section 6.3.6.

9.4 Runaway

A situation when a machine has completely runaway is considered as a catastrophic case. Maximum speed should be computed without any restriction. The designer should foresee only the devices for subsequent closing−down of this machine to prevent an excessively long operation under runaway conditions.

9.5 Response of the governor

The universal characteristic of the regulation process defines the governor's response after a change of load. The equivalent delay time, Δt_e, should be computed taking the maximum velocity of the servomotor piston (i.e. when the spool valve is blocked in the fully open position). The initial values of head, discharge, power etc. should be equal to the rated ones for turbine operation. If the plant has a separate penstock for each machine, Δt_e is computed for one unit only. If several machines use the same penstock Δt_e should be computed for the whole group.

Maximum values of the equivalent delay time Δt_e are usually limited(9) as follows:

(i) for small hydro power plant, which do not regulate frequency or power in the electrical system, Δt_e is not limited,

(ii) for medium size hydro power plants, which make up to 20% of the system's capacity (without exchange of power among electrical systems), Δt_e is 15 to 20 seconds,

(iii) for hydro power plants which cover more than 20% of the system's capacity, the system having reliable interconnections as well as connections with the neighbouring system, Δt_e is 10 to 15 seconds,

(iv) for hydro power plants which maintain frequency in the system or belong to an electric system with weak connections or cover more than 30% of total capacity, Δt_e is less than 8 seconds,

(v) for hydro power plants serving autonomous systems or being connected to isolated networks the delay time is determined according to the customer's demands.

9.6 Waterhammer

There is no specified limit for maximum overpressure resulting from transient conditions, so it has to be determined in each case taking into account technical and economic considerations as well as future operating conditions.

Minimum local pressure in any conduit (tunnel, penstock, pipeline) should not fall below 0.05 MPa (0.5 bar) even temporarily, in order to avoid the danger of water column separation and consequent reverse waterhammer.

9.7 Timing of closure process

A 'perfect' closure process for a simple penstock will provoke a shock equal to[9]

$$\Delta h_{per} = \frac{T_w}{T_f - 0.5\tau} \tag{9.1}$$

where
- Δh_{per} = relative pressure rise related to rated head
- T_w = hydraulic inertia time (s)
- T_f = closing time corresponding to maximum servomotor speed (s)
- τ = $2L/a_w$ (s)
- L = length of the penstock (m)
- a_w = average wave velocity (ms^{-1})

This is an idealised case, and in any real situation the shock would be stronger so the 'perfection coefficient'[9] is introduced as

$$\epsilon_{\Delta h} = \frac{\Delta h_{max}}{\Delta h_{per}} \tag{9.2}$$

where Δh_{max} = relative pressure rise corresponding to the adopted closure process.

For classic hydro power plants (turbine operation only) the above coefficient should not exceed 1.3.

For pumped storage projects the perfect closure process should be defined separately for the cases of power failure (speed not constant) and for the normal stoppage (speed constant). The difference between these two values increases as the specific speed decreases and should be taken into account when designing such plants. At the present time there is no general recommendation.

9.8 Protection against waterhammer

Selection of adequate protection against waterhammer and a good arrangement is the subject of a serious analysis. However at the beginning of the design process (feasibility study) some general recommendations are very helpful, as shown in Fig. 9.1, which refers to installations equipped with Francis type turbines.(9) The notation for the figure is as follows:

T_w = average hydraulic inertia time(s)
T_f = closing time of wicket gates(s)
T_{rp} = closing time of the relief valve(s)
x_{imax} = maximum relative speed of the machine
y_o = maximum relative stroke of the wicket gate servomotor
y_{rp} = maximum relative stroke of the relief valve servomotor.

The diagram covers the following cases:

1. Without protection. If T_w is less than 2 seconds, then maximum speed rise will be within the range $x_{i\,max} < 0.5$. If $2 < T_w < 4$ (seconds) the overspeed will be given by $0.5 < x_{i\,max} < 0.65$. For $T_w > 4$ full runaway should be expected.
2. System protected by a relief valve. If the value of T_w is within 4 to 6 seconds such a measure may be justified. The decision whether to incorporate a relief valve or not is usually based on economic considerations

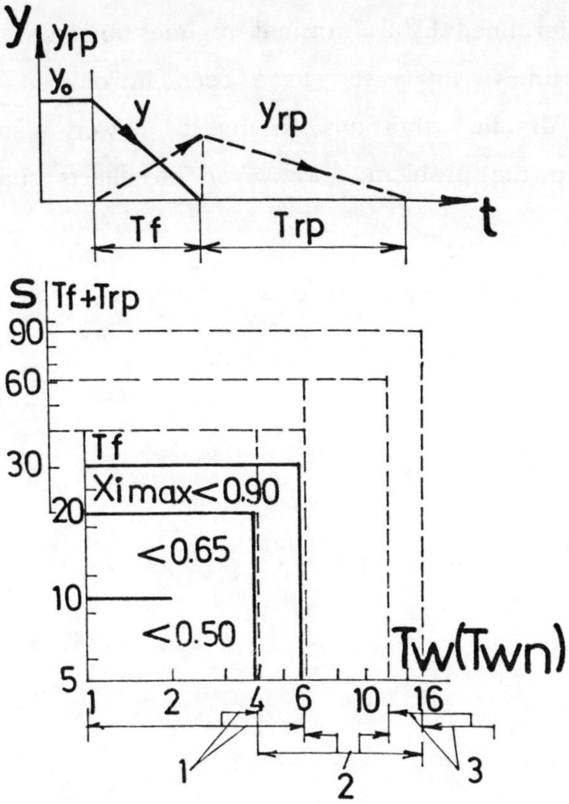

Fig. 9.1 Protection against waterhammer for installation equipped with Francis type turbines

(i.e. too high overspeed may be reduced by adopting shorter closure time which in turn causes higher overpressures and increases the cost of a pipeline. This should be compared with the cost of the valve and all the fittings). For T_W in the range 6 to 12 seconds an economic analysis should decide which is the better solution : to install a relief valve or a surge tank. For $T_W > 16$ seconds, the surge tank is usually necessary.

3. Systems protected by a surge tank. Plants with Pelton type turbines are protected by surge tanks only; jet-deflectors are obligatory parts of the equipment having similar functions as relief valves. The diagram shown in Fig. 9.1 may be used as a guide.

This simple analysis may be used for pumped storage installations as well, but only as an indication, because this problem is much more complex. The final answers should be determined by a transient regimes analysis.

In more recent times, air vessels have been introduced as protection against waterhammer and hydraulic vibrations in hydro power plants, being the most economical solution to the problem. However no charts similar to Fig. 9.1 are yet available.

10 Permitted deviations between computed and guaranteed values

10.1 General

Computations of transient regimes are performed several times during the design process, with various degrees of accuracy and completeness. In many cases only approximate computations are sufficient. As the investigation makes progress, the data and the design become more precise and definite, thus justifying the use of more sophisticated methods of analysis. The question 'how accurate are the results' always hangs in the air. When all the data are available (i.e. machine characteristics, layout, characteristics of tunnels and penstocks, type and shape of surge tanks, timing of wicket gate closure, etc.) the definitive computations are performed. The deviations of the results from these computations from real values may be predicted.

10.2 Machine speed

The maximum machine speed should be computed according to the rules set out in Section 7.2.7 (i.e. $T_{f\ comp} = 1.1\ T_f$). The maximum measured machine speed should be within 2% of the calculated or guaranteed value. This should be taken into account when setting the device which activates protection against machine runaway.

10.3 Extreme pressures

Maximum and minimum pressure should be computed according to Section 7.2.7 (i.e. $T_{f\ comp} = 0.9\ T_f$) and envelopes of pressure drawn for all conduits. Computed or guaranteed values of pressure surges, at both sides of the machine, should be increased by 5%. Further along the conduits the computed values should be increased by an increment (Δh) given by

$$\Delta h = \Delta h_{\ell\ comp} \left[1 + 0.05 \frac{\Delta h_{\ell\ comp}}{\Delta h_{T\ comp}} \right] \tag{10.1}$$

where $\Delta h_{\ell\ comp}$, is the local pressure surge and
$\Delta h_{T\ comp}$ is the corresponding value near the turbine.

This correction takes into account inaccuracies inherent in the data and the numerical procedure.

References

1. Arshenevsky N.N., "Reversible Hydraulic Machines of Pumped Storage Power Plants", (Abratymye hydromachin hydroakumulyrushcih electrostancy) Moskva 1977.

2. Beduci, Giberti, Rocchi et Rovaro, "Interactions between machine valves and regulation equipment in reversible hydroelectrical units", La Houille Blanche, Numero special 7–8, 1980, France, pp. 475–483.

3. Boldy, A.P., "Representation of the Characteristics of Reversible Pump Turbine for Use in Waterhammer Simulations", Proceedings of the 4th International Conference on Pressure Surges, paper G1, pp. 287–296, Bath, England, 21–23 September 1983.

4. Bonin, C.C., "Waterhammer Damage to Oigawa Power Station", Journal of the Engineering for Power, Transactions of the ASME, August 1960.

5. Chaudry, H.M., "Applied Hydraulic Transients", VNR, New York, 1979.

6. "Closed – Conduit Flow", editors Yevdjevitch, V. and Chaudry, H.M., Water Resources Publications, 1981.

7. Glass, W.L., "Cavitation and Corrosion in a Pumping Main", Proceedings of the Third International Conference on Pressure Surges, paper H3, pp. 401 – 414, Canterbury, England, 25–27 March 1980.

8. Grujic, Lj., "Automatic Control of Hydroelectric Power Plants" (Automatsko

upravljanje hidroenergetskih postrojenja), Faculty of Mechanical Engineering, Beograd, 1980.

9. Guidelines for Analysis of Regulation of Hydroelectric Power Plants, (Rukavodstvo po proyektirovanyu technologicheskyh rezhimov regulirovanya hydroelectrostanzy) Moskva 1977.

10. International code for testing of speed governing systems for hydraulic turbines, IEC Publication 308, Geneve, 1970.

11. Krivchenko, G.I., Arshenevsky, N.N., Kvyatkovskaya, E.V., Klabukov, V.M., "Hydraulic Transients in Hydroelectric Power Plants", (Hydromechnicheskye perechodnye processy v hydroenergetycheskyh ustanovkah), Moskva, 1975.

12. Krivchenko, G.I., Arshenevsky, N.N., Klabukov, V.M., "Regulation of Kaplan Turbines", (Rezhimy regulirovaniya pavarotnolopastnih hydroturbin) Moskva, 1960.

13. Krivchenko, G.I., Kvyatkovskaya, E.V., Matveyev V.N., Leontev A.I., Kossolapov, T.V., "Safe Operation of Hydrostation Units with Kaplan Turbines", (K vaprosu bezopasnosoy raboty agregatov hydroelectricheskyh stanzy pavarotno lapastnymy turbyny), Hydrotech. Construct., V.18, No.3, March 1984, pp. 119−123.

14. Martin, C.S., "Stability of Pump Turbines during Transient Operation", Proceedings of the 5th International Conference on Pressure Surges, paper C3, pp. 61−71, Hannover, F.R. Germany, 22−24 September, 1986.

15. Pejovic, S., "Guidelines to Hydraulic Transients Calculations in Hydro Power Installations", (Preporuke za proracune prelaznih procesa u hidroenergetskim postrojenjima), Beograd, 1984.

16. Pejovic, S., "Water Hammer and Hydraulic Transients", (Hidraulicki udar i prelazni rezimi hidropostrojenja) Beograd, 1977.

17. Pejovic, S., "Reverse Water Hammer and Incident of Hydro Power Plant "Zvornik" ", (Povratni hidraulicki udar i havarija agregata HE "Zvornik", Beograd, 1978.

18. Pejovic, S., "Hydraulic Vibrations of Hydro Power Plants", (Hidraulicne

vibracije hidropostrojenja), Beograd, 1978.

19. Pejovic, S., Krsmanovic, Lj., Jemcov, R., Crnkovic, P., "Unstable Operation of Highhead Reversible Pump Turbines", IAHR Symposium, Leningrad, 1976.

20. Richtlinien fur die Erstellungen von stahleren Druck–rohrleitungen fur Wasserkraftanlagen VWEW, Frankfurt, 1966.

21. Vladisavlev, L.A., "Vibrations of the Units in Hydroelectric Power Plants", (Vibrazya hydroagregatov hydroelectricheskyh stanzy), Moskva, 1972.

22. Welded – Steel Penstocks, Bureau of Reclamation, Washington, 1977.

23. Wylie, B.E., Streeter, L.V., "Fluid Transients", McGraw–Hill, New York, 1978.

Appendix I Additional data relating to the governor

I.1 Introduction

Governor characteristics influencing the analysis of hydraulic transients and hydraulic oscillations within the system are summarised in the following sections.

I.2 Description of the governor

A governor for turbomachines with fixed blades (Francis, Pelton) is shown schematically in Fig. I.1. It consists basically of three elements:

(i) regulation block with sensors, signal analysers, control and regulating devices.

(ii) execution block comprising servomotors for opening or closing of flow–regulating devices (wicket gates, valves, needle valves and gates, etc.).

(iii) feedback control.

The input variable is the dimensionless change of runner speed (x_i), defined as

$$x_i = \frac{n - n_o}{n_o} \tag{I.1}$$

Two command signals are introduced.

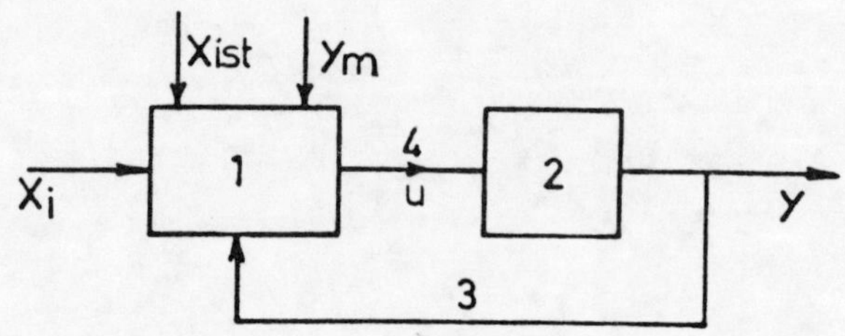

Fig. I.1 Block diagram of a governor.
1: Regulation block, 2: execution block,
2: feedback control, 4: regulating action

(i) x_{ist} (−) dimensionless manual speed setting.

(ii) y_m (−) limit of dimensionless servomotor piston extreme position.

The output variable of the regulation block is the dimensionless displacement of the spool valve (u), given by

$$u = \frac{x_r}{x_{rR}} \tag{I.2}$$

and is also the input variable for the execution block which controls the servomotor. The dimensionless displacement (y) of the servomotor piston, defined as

$$y = \frac{Y - Y_m}{Y_m} \tag{I.3}$$

is the output variable of the second block. Feedback control provides stability of the regulation process and is, therefore, necessary.

This is the simplest scheme of a governor which covers the most common types. More detailed descriptions of governors can be found in the specialised literature and obtained from appropriate manufacturers.

The governor characteristic relevant to the analysis of transient regimes is the permanent speed droop (b_p). This is the relation between relative deviation of a servomotor stroke, y, and relative change of speed, x_i, shown in Fig. I.2. The value of b_p is given by the slope of the line shown in Fig. I.2. It can be adjusted in a range from zero to 0.1, and is usually in the range 0.004 to 0.06[8,9,10]. This characteristic of a governor could be changed on site. By turning the mechanism for manual speed setting, the speed droop line will be displayed, see Fig. I.2, so the servomotor stroke of the wicket gates (y) will be different for the same input signal x_i. The properties of turbomachines regarding load acceptance are thus considerably changed and this fact must be taken into account in the analysis of transient regimes.

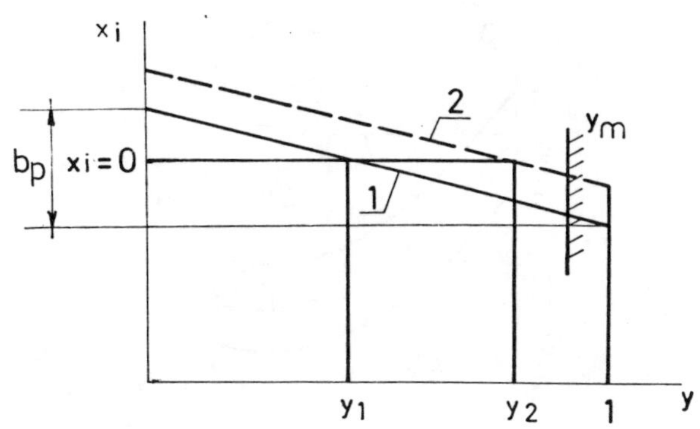

Fig. I.2 Speed–droop relationship

The maximum stroke of the servomotor is limited to the value y_m, see Fig. I.2. This value can also be changed on site. Setting $y_m = 0$, one actually closes down the wicket gates and stops the unit.

I.3 Characteristics of the servomotor

The speed of closing–down or opening–up of the wicket gates (needle valve in Pelton units) has a profound influence on pressure surges in the system. This speed is determined by servomotor piston movements which must therefore be simulated very accurately.

The relevant factors are

(i) external forces which the servomotor has to overcome,

(ii) oil pressure in the regulating system,

(iii) displacement of the spool valve.

External forces are hydraulic forces acting upon the wicket gates (needle valve) plus friction forces. The force on the piston, which is greater than all the external forces, is proportional to the difference between pressures acting on both sides (Δp). This difference depends on the relative position (y) of a piston within the cylinder and on the kinematics of the mechanism. The relation of $\Delta p = f(y)$ for a typical Francis type turbine is shown in Fig. I.3. The broken line represents the hydraulic forces on the wicket gates and the full lines represents the sum of external forces, friction forces being added for opening−up or subtracted for closing−down.

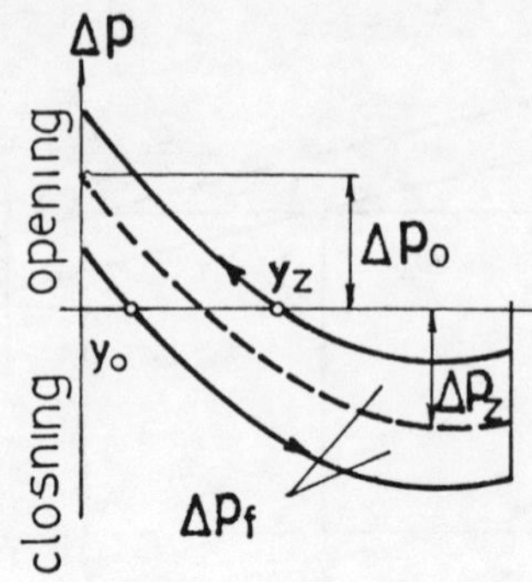

Fig. I.3 Wicket gates servomotor differential pressure against wicket gates opening

Oil pressure in the regulating system could suddenly be lost due to equipment breakdown or human error. In that case, the wicket gates will assume either the position y_z, if the unit was operating beforehand, or the position y_o, if the unit was closed down (see Fig. I.3), allowing water to flow through the machine. Hydraulic forces are not only functions of wicket gate opening (related to servomotor stroke y), but depend on turbine heads also. Friction forces depend on head, the quality of manufacture and the maintenance of the mechanism. Therefore, the pressure increment, Δp, deviates from the average values given in Fig. I.3. For good systems, properly maintained, the deviations should not exceed the following limits:

(i) deviation of pressure for overcoming hydraulic forces

$$\Delta p_1 \leqslant 0.1\ (|\Delta p_o| + |\Delta p_z|) \qquad (I.4)$$

where Δp_o and Δp_z are maximum pressure differences necessary to open–up and close–down the wicket gates[12].

(ii) deviation of pressure for overcoming friction forces

$$\Delta p_2 < 0.2\ |\Delta p_{f\ max}| \qquad (I.5)$$

where $\Delta p_{f\ max}$ is the maximum friction force in the mechanism.

Fig. I.4 shows the range of allowable pressure differences in the servomotor, related to the servomotor stroke.

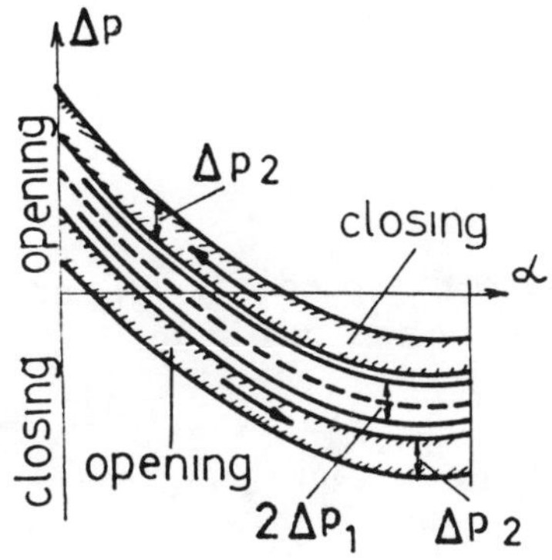

Fig. I.4 Ranges of allowable variation of servomotor pressure

I.4 Minimum oil pressure

The minimum oil pressure (p_{min}) in the reservoir of the regulating system, at the end of a full cycle, has to be at least

$$p_{min} > k\ p_s \qquad (I.6)$$

where p_s is a design pressure (used for determination of a piston diameter) and k is a safety factor (1.3 to 1.4)[12].

I.5 Speed characteristic of a servomotor

Servomotor piston speed is usually determined for normal pressure in the oil reservoir[9]. It depends on displacement (u) of the spool valve distributor which is the command signal. A typical relation between the relative servomotor speed, dy/dt, and u is shown in Fig. I.5. Three distinct zones are visible on this diagram:

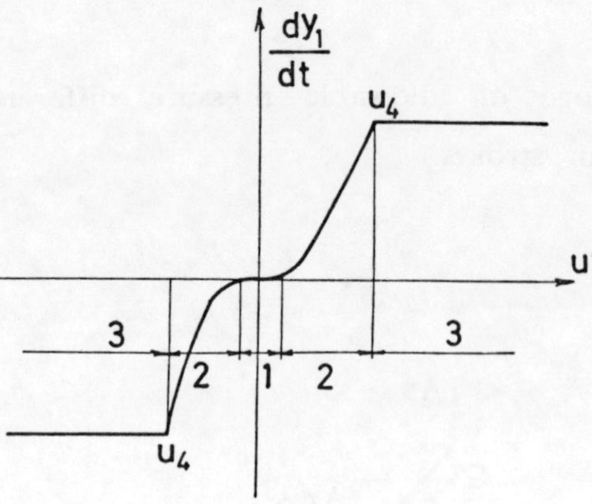

Fig. I.5 Relationship between speed (dy_1/dt) and displacement (u) of the distributor.

(i) zone of insensitivity (1), for $|u| \leqslant 0.0002$, which roughly corresponds to the angular speed change of $|x_i| \leqslant 0.02\%$.

(ii) zone of changing servomotor speed (2), for $|u| \leqslant 0.01$ to 0.02 or $|x_i| = 1$ to 2%.

(iii) zone of maximum servomotor speed (3), for $|u| \geqslant 0.01$ to 0.02 or $|x_i| \geqslant 1$ to 2%.

Zones 1 and 2 are important when analysing the stability of the regulating system for various initial conditions, under small disturbances, with the purpose of finding out the best characteristics of the governor.

Zone 3 corresponds to the maximum servomotor speed, with the ports of the spool valve fully open. This is very important in the analysis of transient regimes where major disturbances are being simulated.

Appendix II Advanced studies of transient regimes

II.1 Introduction

The scope of analysis given in the main text is usually sufficient to obtain a complete answer regarding the safety of a plant. The analysis of several normal emergency and catastrophic cases will enable the designer to select proper methods for protection for the plant against possible damage due to transient regimes.

However the research in this area is not complete and there are still unresolved problems such as:

(i) functionability of the plant in transient regimes,
(ii) three–dimensional unsteady flow in the turbine,
(iii) fluid–structure interaction during transients,
(iv) two–phase flow in the system.

II.2 Functionability of the plant

A hydro power plant is a very complex system with several hydraulic and electric machines with corresponding excitation and governing systems. The real boundary of this system is not at the machine shafts, but rather at the connection to the electrical network or even at the electrical energy users. Disturbances, causing transients, may come from either the electric or the hydraulic side of the system, or both. When steady state no longer exists, the currents in a synchronous electric machine change, the speed of rotation varies and the governor steps in commanding the wicket gates to close–down or open–up. A new steady state will be reached, asymptotically or in an oscillatory manner.

To describe and analyse this process in full a very detailed mathematical model is necessary and requires the following additional data:

(i) layout of the electrical installations with the plant,

(ii) data on synchronous machines, generators and motors,

(iii) data or exciters and transformers.

(iv) data on energy users.

Such models are being developed(24) but are not yet in common use. Such a model, when completed and fully calibrated, might help in analysing and circumventing the potential operational difficulties of unstable operation, too strong pressure surges and unwanted interactions between the operating units.

II.3 Three–dimensional unsteady flow in turbines

One of the major deficiencies of mathematical modelling of transient regimes is the simplification of turbines and pump turbines. They are usually represented as a node (i.e. as a point) described by characteristic curves relating discharge Q, head H, speed ω, torque M as functions of wicket gate opening a_o. This is a static, one–dimensional representation of a dynamic three–dimensional hydraulic machine. Attempts to overcome this situation are being made, both experimentally and theoretically.

Unevenness of the flow field below a turbine runner was experimentally studied in the U.S.S.R(25). For all standard machines the diagrams showing variations of flow and pressure under the runner for loads ranging from zero to full load were prepared. These are particularly useful in the analysis of reverse waterhammer.

Yamabe[26] has studied the hysteresis characteristics of pump turbines with fixed blades, a fact which is not, to the authors' knowledge, incorporated in the analysis of transient regimes.

Kinematic analysis of wicket gates has, in rare instances, been analysed by researchers working for the manufacturer[27].

Hydrodynamic forces acting on wicket gates and runner blades of modern pump turbines have been measured and analysed on small–scale models and prototypes, both for normal and abnormal operating regimes. Particular attention was given to regimes producing excessive head or torque vibrations and the findings plotted on regular four–quadrant performance characteristic diagrams, in the form of lines of iso–strength of vibrations[28].

The results of experimental investigations have helped in the formulation of mathematical models of dynamic behaviour of turbines[29]. The aim of these efforts is to create a model which can account for both the steady–state characteristics of machines and for hydraulic oscillations of flow and pressure which are generated during normal operation.

II.4 Dynamic interactions between a structure and fluid

This problem is particularly important when analysing the effects of earthquakes upon non–rigid structures like pipe networks, bottom outlets of a dam, tunnels, etc. Several simplified models of a moving structure with fluid in it were proposed, but the problem calls for further research[29].

II.5 Two–phase flow

Two–phase flow in a hydro power plant is not a frequent event. However, in some low head plants pressure might drop well below atmospheric pressures, releasing dissolved air (in a draft tube for instance). The mixture of free air, vapour and water has very different physical characteristics from a homogeneous water body. The most important difference, from the point of view of transient regimes, is that the wave speed in this mixture falls very low, possibly in the range 30 to 50 ms^{-1}.

II.6 Hydrodynamic forces

Hydrodynamic forces acting on valves during transient regimes have to be measured on a model or on a prototype in order to make it possible to undertake a detailed analysis of closing or opening procedures. Data on servomotors and the governing system, as well as on the kinematics of the mechanism, should be supplied by the designer or the manufacturer.

II.7 Miscellaneous problems

Other unresolved problems are:

- three-dimensional flow in/out of surge tanks of irregular shape (danger of entrapping huge quantities of air);
- effects of unsymmetric branching on pressure surges;
- exact modelling of friction losses in near-zero flows.

II.8 References

24. Smith J.R., McLean R., Robbie J.F. "Assessment of hydroturbine models for power-system studies". Proc. I.E.E., Vol. 130, No.1, Jan. 1983, pp. 1-7.

25. Krivchenko, G.I., Lectures held at the Faculty of Mechanical Engineering, Belgrade 1978.

26. Yamabe M. "Hysteresis characteristics of Francis pump turbines when operated as a turbine". Trans. ASME. Jnl. of Basic Engineering, Vol. 93, No.2, March 1971, pp. 80-84.

Yamabe M. "Improvement of hysteresis characteristics of Francis pump turbines when operated as a turbine", Trans. ASME. Jnl. of

Basic Engineering, Vol. 94, No. 5, Sept. 1972, pp. 581–585.

27. Jordan V., "Kinematische berechung des regulier mechanismus einer Wasser–Turbine mit Hilfe des elektronischen Rechner", Comm. Third World Congress Theory of Machines and Mechanisms, Kupari, Yugoslavia, Sept. 1971, Vol. H, Paper H14.

28. Casacci S., Wegner M., Guiton P., "Analyse des vibrations et des instabilities de functionnement dans les turbines–pumpes", La Houille Blanche, No.6, 1971, pp. 519–530.

29. Fanelli M. "Further considerations on the dynamic behaviour of hydraulic turbomachinery", Waterpower, June 1972, pp. 208–222.

Appendix III Typical transient regimes in hydro power plants

The scope of analysis for a feasibility study or a detailed design is as follows.

III.1. Hydro power plant with Kaplan type turbines:

III.1.1 *Normal operating conditions:*

(i) Start–up and loading of turbines;

(ii) Change of load (i.e., change of wicket gates and/or runner blades angle) – computation of the equivalent delay time;

(iii) Shift to compensation operating regime;

(iv) Normal stopping of machines;

(v) Load rejection and idle rotating;

(vi) Power failure with rapid closing down;

(vii) Stability of regulation.

III.1.2 *Emergency operating conditions:*

(i) Rapid closure without cushioning at small openings;

(ii) Sudden closure of the inlet gate;

(iii) Stoppage after a complete outage (the units go to runaway).

III.1.3 *Catastrophic cases:*

(i) Full runaway conditions;
(ii) Sudden closure of wicket gates or a major valve;
(iii) Reverse water hammer;
(iv) Oscillatory movements of blades with a resonant frequency;
(v) Earthquake.

III.2. Hydro power plant with Francis type turbines:

III.2.1 *Normal operating conditions:*

(i) Start−up and loading of turbines;
(ii) Shift to compensating regime;
(iii) Normal stopping;
(iv) Change of load − computation of the equivalent delay times;
(v) Load rejection and idle rotating;
(vi) Power failure with rapid closing down;
(vii) Hydraulic vibrations.

III.2.2 *Emergency operating conditions:*

(i) Rapid closure without cushioning in one unit;
(ii) Failure of one pressure regulating device;
(iii) Closure of the main valve in penstock;
(iv) Protection against runaway;
(v) Closure of the inlet gate.

III.2.3 *Catastrophic cases:*

(i) Rapid closure (without cushioning effect) on two or more units;
(ii) Failure of two or more pressure regulating devices;
(iii) Full runaway;
(iv) Sudden closure of wicket gates;
(v) Sudden closure of main valves or gates;
(vi) Reverse water hammer;
(vii) Oscillations of wicket gate blades;

(viii) Auto–oscillations;

(ix) Earthquake

III.3. Hydro power plants with Pelton type turbines:

III.3.1 *Normal operating conditions:*

(i) Start–up and loading of turbines;

(ii) Shift to compensating regime;

(iii) Normal stopping;

(iv) Change of load – computation of the equivalent delay times;

(v) Load rejection and idle rotating;

(vi) Power failure followed by rapid closure of turbine needle valves;

(vii) Hydraulic vibrations.

III.3.2 *Emergency operating regimes:*

(i) Rapid closure without cushioning effect on one unit;

(ii) Failure of one flow deflector;

(iii) Closure of one turbine valve (its needle had failed to close the pipe);

(iv) Closure of the inlet gate;

(v) Protection against runaway conditions;

III.3.3 *Catastrophic cases:*

(i) Rapid closure without cushioning of several units;

(ii) Failure of two or more flow deflectors;

(iii) Full runaway;

(iv) Sudden closure of all needles;

(v) Sudden closure of penstock valves;

(vi) Oscillatory movements of turbine needles;

(vii) Auto–oscillations;

(viii) Earthquake.

III.4. Pump operating regimes of hydro power plants with Kaplan type reversible units (pump–turbines):

III.4.1 *Normal operating conditions:*

(i) Start–up to pumping mode;

(ii) Stoppage from pumping mode;

(iii) Shift from pumping to compensating regime;

(iv) Shift from compensating regime to pumping mode of operation;

(v) Shift from pumping to turbining regimes;

(vi) Load rejection in pumping operating regimes with rapid closure of wicket gates.

III.4.2 *Emergency operating regimes:*

(i) Rapid closure without cushioning effect;

(ii) Closing down of the inlet gate;

(iii) Protection against runaway;

III.4.3 *Catastrophic cases:*

(i) Full runaway;

(ii) Sudden closure of wicket gates;

(iii) Sudden closure of inlet or outlet valves;

(iv) Earthquake.

III.5. Pump operating regimes of hydro power plants with Francis type reversible units (pump–turbines):

III.5.1 *Normal operating conditions:*

(i) Start–up to pumping operation;

(ii) Stoppage from pump operation;

(iii) Shift from pumping to compensating regimes;

(iv) Shift from compensating to pumping regimes;

(v) Shift from pumping to turbining (power generating) regimes;

(vi) Load rejection in pumping operation and rapid closure of wicket gates.

III.5.2 *Emergency operating regimes:*

(i) Rapid closure – without cushioning effect – of one unit;

(ii) Closure of the inlet gate;
(iii) Protection against runaway;
(iv) Closure of the valves in penstock.

III.5.3 *Catastrophic cases:*

(i) Rapid closure, without cushioning, in two or more units;
(ii) Full runaway;
(iii) Sudden closure of wicket gates (slamming);
(iv) Sudden closure of valves in the penstock (slamming);
(v) Earthquake.

III.6. Safety

Safety coefficients: K_1 (related to the ultimate strength of material) and K_2 (related to the yield limit) should have the following values:

		K_1	K_2
1.	Normal operation	4 to 5	1.5 to 2.2
2.	Emergency conditions	2 to 3	1.4 to 2.0
3.	Catastrophic cases	1	1

The plant is not usually designed to withstand the catastrophic events because the costs will be prohibitive. If some damage happens to the plant in catastrophic circumstances it is regarded to as a 'vis major' case.

III.7 Important warning:

The list given above is rather general. This list should be compiled again in each particular case, taking into account specific features of the plant in question and requests of all concerned. Special care should be devoted to the small power plants because there is a strong tendency to reduce the costs of analysis and design to the minimum. The transient regimes analysis is sometimes omitted altogether for the same reason. This practice is dangerous, and it is much better to entrust this simplified and reduced analysis to a well−trained and experienced specialist who can help to prevent the incidents in future plant even when working within the narrow constraints of time and money.

Index